Fish Nutrition and Feed Technology

A Teaching Manual

Fish Nutrition and Feed Technology
A Teaching Manual

S. Athithan
N. Felix
N. Venkatasamy
Fisheries College & Research Institute
Thoothukudi – 628 008,
T.N., India

2024
Daya Publishing House®
A Division of
Astral International Pvt. Ltd.
New Delhi - 110 002

First Published, 2012
Reprinted, 2024

ISBN: 978-81-7035-781-0 (Hardbound)

Published by : **Daya Publishing House®**
A Division of
Astral International Pvt. Ltd.
– ISO 9001:2015 Certified Company –
4736/23, Ansari Road, Darya Ganj
New Delhi-110 002
Ph. 011-43549197, 23278134
E-mail: info@astralint.com
Website: www.astralint.com

Printed at : **Replika Press Pvt. Ltd.**

Tamil Nadu Veterinary Animal Sciences University

Dr. R. Prabaharan, Ph.D.,
Vice-Chancellor

Madhavaram Milk Colony
Chennai – 600 051

Foreword

Fish nutrition is fundamental to the most of the aquaculture practices in the world. Although research support aims, in part, to promote the culture of species requiring no supplementary feeding, it is recognized that formulated feeds are required to increase productivity of many important species now being cultured within the region. This requirement will undoubtedly continue in the future. The present publication which deals with all the aspects of fish nutrition and feed technology is a timely one to fill up this long felt need. I am confident that the chapters discussed by the authors in this manual will be of immense value to the students and researchers who would like to pursue research in fish nutrition. The authors' hard work, dedication and passion for the benefit of the students is highly appreciated.

R. Prabaharan

Preface

Aquaculture is now recognized as a viable and profitable enterprise worldwide. As aquaculture technology has evolved, the push toward higher yields and faster growth has involved the enhancement or replacement of natural foods with prepared diets. In many aquaculture operations today, feed accounts for more than a half of the variable operating cost. Therefore, knowledge of nutrition and practical feeding of fish is essential for successful aquaculture. This teaching manual is not written exclusively only for scientists but also for students who are practicing nutritionists and aquaculturists. The first part of teaching manual covers on types of feed, physiology of digestion and metabolism in fishes, food and feeding habits of fish. The second part includes nutritional requirements of cultivable fin and shellfishes – proteins, fats, carbohydrates, vitamins & minerals and nutritional bioenergetics of fish. Aquafeed or diet formulation, micro feeds or micro particulate feeds, feed ingredients and feed additives are discussed under the third part. Physical chemical, biological and microbiological evaluation of feeds & feed ingredients were presented under the fourth part. The main fifth part includes feeding strategies, feed processing, feed handling and storage. The last part of this manual is dealing with antinutritional factors in aqua feeds, nutritional diseases, probiotics and their use, economic aspects with regard to use of aqua feed and feed management in aquaculture. Each part has been meticulously prepared and fortified with new information. Nonetheless, the material presented has been thoughtfully selected and updated to make it of maximum use to persons whose interests range from general aquaculture to animal nutrition to feed manufacture and student communities in particular.

S. Athithan

N. Felix

N. Venkatasamy

Contents

Chapter 1

Introduction

The word 'nutrition' denotes the food or dietary needs of the body or act or processing of nourishing (supply food for survival and growth). Fish nutrition is defined as, "studying the nature of fish food; dietary needs or nutrient requirement of the fish body". Based on the (*i*) dietary needs of the fish body (nutrient requirement), (*ii*) physiology of digestion and (*iii*) food and feeding habits of fish, one can develop a feed conforming scientific formulation of feed, optimal processed feed and supply of feed judiciously to fish.

Farming systems followed both in finfish and shellfish are of traditional, extensive, modified extensive, semi intensive, intensive/ super intensive types. Feed is the major operational input in aquacultural practice. In any aquacultural practice, feed cost constitutes the range of 30–70 per cent or even up to 80 per cent depending upon the type of farming (Carp farming 30–50 per cent; Shrimp farming 50–80 per cent).

Importance

Capture fishery production India, both in the marine as well as freshwater sector, is being increasingly plateaued. While the annual growth rate of capture fishery has been somewhat stagnant, aquaculture production has been recording a phenomenal annual growth of 9 per cent. This is because, over the years, aquaculture has become remarkably significant, making it one of the fast expanding bio industries. Nevertheless, the global demand for fish exceeds the production and the per capita fish consumption is steadily increasing. The rising living standard in developing countries and the increase in world population are set to create greater demand for fish in future. This will increase the demand for farmed fish, considering that the scope for augmenting capture fish production, both marine and inland, that is seen to be bleak now. For farmed fish production to increase, several technological constraints would have to be overcome. Insufficient availability of natural food and shortage of nutrition oriented supplementary feeds, consequential slower growth rate, limitations on broodstock availability and quality seed production, loss due to deficiency and diseases as well as environment concerns are some of these constraints. Modernization of aquaculture and greater understanding of nutritional requirements, digestive system, growth factors, reproductive physiology and larval development of culture species as well as improvement of growth potential through genetic alterations open up far greater scope and challenge for fish nutritionists in the development and diversification of aquaculture feeds.

The feed supplies the dietary requirements for the organisms. The growth of the fish directly depends on the feed. So, the feed must provide all the nutrition to the organisms. The factors which affect feed design are depending upon the cultured species, size and stage of life cycle, reproduction stage of life cycle, system of culture, feeding habit, environmental condition, stocking density and size and shape of feed.

Challenges ahead in Fish Nutrition and Feed research

Substitute for fish meal, antinutritional factors, Growth promoters, Broodstock nutrition, Larval nutrition (formulated larval diets, live feeds, bioenriched feeds), Health management via feeds, Stability of feeds, Finishing diets, Attractants in feeds, Organoleptic improvement, Locally made feeds are some of the challenges ahead in fish nutrition and feed research.

Self Assessment Questions

1. What are the different meanings of the word "Nutrition"?
2. What is fish nutrition?
3. What is Aquaculture nutrition?

4. Differentiate the term between fish in nutrition and fish nutrition.
5. In what way, the term fish nutrition is differed from aquaculture nutrition.
6. What are the primary aspects to be studied before making feeds?
7. What are the factors affecting feed design, production and feeding?
8. What are the different challenges ahead in fish nutrition and feed research?

Chapter 2
Types of Feeds Used in Aquaculture

Successful production good quality seed and table size fish can be achieved, by and large, through the operation of 2 regimes, *viz.*, feeding the larvae and fish with nutritionally balanced feed; maintain the water quality parameters in optimum range. The feed is differentiated by its physical properties and its contents. There are many varieties of feed used in aquaculture system. The feed used by the farmer is according to their convenience. The wrong usage of feed must be affecting the aquaculture system. The feed is broadly classified into two types such as natural and artificial feed.

I. Natural Feed

It is otherwise called as viable feed or live feed. Natural feeds generally serve as living capsules of nutrition. They are live feed or naturally occurring substances which can perform the growth prominently. These feeds are very cheap and occur very easily. Live feeds do not have most of the disadvantages of artificial feeds. Natural feeds are further classified as aquatic and non-aquatic live feeds. The examples for aquatic live feeds are:

Unicellular Algae

These are microscopic single-celled plants which are the major primary producers, photosynthesing food from the pond nutrients utilizing light energy. They reproduce very quickly if given a good supply of nutrients.

Filamentous Algae

These plants also photosynthesize in the same way as unicellular algae. They consist of colonies of small algal cells attached together. Often regarded as relatively inedible and therefore wasteful in ponds for some aquaculture species, but the growth of some types (the green filamentous algae) is actively encouraged for shrimp and milkfish culture in some countries. Filamentous algal colonies can be very large and may form dense masses causing physical problems in pond management.

Lab-Lab

A community consisting mainly of benthic (bottom dwelling) blue-green algae and diatoms, together with other plants and animals. It is a favourite food for milk fish.

Lumut

A community consisting mainly of filamentous green algae, together with other algae and animals. It is a favourite food for mullet.

Bacteria and Fungi

Microbial organisms which grow very rapidly and live on detritus at the bottom of the pond–dead phytoplankton cells, leaves, and dead animal tissue.

Zooplankton

Whereas the microscopic plants are referred to as phytoplankton, zooplankton is a term which refers to microscopic animals. Zooplankton and phytoplankton are together referred to by the general name plankton. The zooplankton include rotifers (known as wheel animalcules), of which *Brachionus plicatilis* is the best known example, which eat algal cells; cladocerans, which include the water fleas, of which *Daphnia* spp. and *Moina* spp. are the best known examples; copepods like *Calanus* spp. and *Cyclops* spp. and anostracans are commonly known as 'fairy shrimp'.

The examples for non-aquatic live feeds are:

Mud Eaters

Animals like insect larvae such as chironomids and various worms, which ingest mud and detritus and derive nutrient from the bacteria and fungi living on it, belong to this category.

Other Insect Larvae

Living in water, these larvae prey on other food animals, as well as the fry of shrimp and cultured fish, but themselves form a food for larger animals.

II. Artificial Feed

These are otherwise called as compound feeds. This type of feed is produced by man using available raw materials and medicine. These feed are produced based on dietary formulation specific to individual fish species. Artificial feeds are again classified as purified or semi-purified diets and practical diets.

Purified or semi-purified feeds are mainly used at laboratory level for quantification of nutrient requirement for many cultivable fishes. The ingredients used are purified and refined one. The common protein source used in semi purified diet is gelatin–casein (90 per cent), carbohydrate source is corn starch, lipid source is corn oil and binder source is carboxyl methyl cellulose (CMC).

Practical diets are mainly used at farm level and the ingredients used are not purified and refined one. Practical diets are further classified based on (*a*) stages of the life cycle, (*b*) forms or moisture content and (*c*) uses at farming level.

Based on their Stage of the Life Cycle

1. Starter feed, crumble feed or larval feeds or fry feeds (minced diets, micro feeds – MED, MBD and MCD, egg custard, live feeds)
2. Fingerling feeds (Grower)
3. Grow out feeds (Finisher)

4. Brood stock feeds
5. Product quality feeds (specific purpose food) pigmented and medicated

Usually, the fish might be changing the feed habit according to the stage. So, feed must provide the nutrition according to the life stage. Hence, they are classified as starter or larval feed and grower (fingerling stage). Some special type of feeds called micro feed is also used. These feed supplies the equal dietary compound as micro particulate, microencapsulated, micro bounded and micro coated.

Based on their Moisture Content or Forms

According to the percentage of water content, they are differentiated into dry feed and non dry feed. Both the dry feed and non dry feeds have got their own merits and demerits. Dry feed is used more commonly than the other type.

Dry feed (7–13 per cent moisture) has 2 types namely, floating and sinking feeds. Floating feed is used for surface and column feeding fish (Catla, silver carp, sea bass, grouper, trout, tilapia, ornamental fishes).

Non dry feed has 3 types, namely, semi-moist feeds (15–25 per cent moisture), moist feeds (26–45 per cent moisture) and wet feed (46–70 per cent moisture).

Based on their Uses in Farming Practice

In aquaculture industry, different types of feed are used in different farming systems. Intensive farming culture organisms require more dietary nutrition than any other extensive or semi intensive system. This feed is divided into two namely the feed which is supplement with the natural food source is called supplementary feed (traditional, extensive, modified extensive, semi-intensive) and supply all the major and micro nutrient to compensate their requirement is called complete feed (intensive, super intensive). Complete food supply all energy requirement, all gross major nutrient requirement and all micro nutrients (lesser and minute). Supplementary feed supplement the natural food sources which supply major nutrients only.

Natural Feed	Artificial Feed
Advantages	*Advantages*
Stabilize the water quality through absorption of inorganic minerals & other mechanism	More fish production Supply all essential energy requirement to fish
Provide oxygen to the system by photosynthesis	Supply gross major nutrient requirements
Reduce the toxic effect of NH_3 Inhibit the development of benthic and filamentous algae	Supply minor nutrients (micro)
Compete with bacteria and may decrease the possibility and frequency on the occurrence of the disease	
Serve as environmental indicator	
Disadvantages	*Disadvantages*
After initial growth, it needs supplementary feed or complete feed for better growth.	Chance for aesthetic appearance since it contains muddy odour, pigments, hormones etc.
Less fish production	Water quality to be maintained through out culture period with utmost care
	Depuration must be done

Dry Feeds	Non Dry Feeds
Advantages	*Advantages*
Moisture (7–13 per cent)	Semi moist feeds (15–25 per cent)
Easy to transport, manufacture and store	Moist feeds (26–45 per cent)
	Wet feed (46–70 per cent)
Produced to suit specific needs	More palatable because of soft consistency
Permit production of specialized feeds such as medicated, pigmented feeds	

Dry Feeds	*Non Dry Feeds*
Disadvantages	*Disadvantages*
Less palatable because of hard consistency	Transportation and storage under refrigeration pose problem
	Irregular availability of raw fish in adequate quantities
	Chance for pathogen introduction if not cooked properly
	Unconsumed feeds may affect water quality

Development of Natural Food in Ponds

Example: Composite Fish Culture

Organic fertilizer

↓

Cow dung @ 10 t/ha/yr

↓

Applied in 10 installments

↓

1000 kg – Ist dose

↓

Apply every 15 days

Inorganic fertilizer

↓

Urea → 200kg/ha/yr

Super phosphate → 250 kg/ha/yr

Muriate of potash → 40kg/ha/yr

↓

Each divided into 10 portions and then and applied

↓

N → 20 kg/ha–Ist dose

P → 25 kg/ha–Ist dose

K → 4 kg/ha–Ist dose

↓

Apply every 15 days

↘ ↙

Alternating with organic and Inorganic fertilizers

Self Assessment Questions

1. Differentiate natural vs. artificial feeds.
2. Why live feeds are considered as "living capsules of nutrition"?
3. What is the favorite food for milk fish?
4. What is the favorite food for mullet fish?
5. What are instant algae?
6. How the feeds are classified?
7. Compare practical and purified diets.
8. Merits and demerits of natural and artificial feeds.
9. Advantages and disadvantages of dry and non-dry feeds.
10. How natural food is propagated in composite fish culture practice?
11. How natural food is propagated in shrimp culture practice?
12. How natural food is propagated in scampi culture practice?
13. How natural food is propagated in composite carp culture practice?
14. Collect name of different species of phyto and zooplankton in freshwater.
15. Collect name of different species of phyto and zooplankton in seawater.
16. Collect names of different species of non-aquatic live feeds.
17. What is depuration? Why it is done for harvested freshwater fishes that to carps?
18. Differentiate supplementary and complete feed.

Chapter 3

Physiology of Digestion and Metabolism in Fishes

In fish and shellfish being the cold blooded animals, the digestion process is somewhat different when compared to the terrestrial animals. Similarly, the mechanism of digestion and absorption process is quite different in finfishes and shellfishes. The details of the physiology of digestion in finfishes and shellfishes are given below.

What is Digestion?

The basic function of digestive system is to dissolve foods by rendering them soluble so that they can be absorbed and utilized in the metabolic process. The system may also function to remove dangerous toxic properties of certain food substances.

Digestion (Absorption) – passage of food through lining of digestive tract in to bloodstream

Mouth (Ingested food) ——————→ **Anus** (faecal matter)

Ingested food $\xrightarrow{\text{(enzymes)}}$ simple, small and absorbable or soluble $\xrightarrow{\text{(gut wall)}}$

Blood stream ——————→ Metabolic process

Physiology of Digestion in Finfishes

Movement of Food in the Tract

It is similar to that of higher vertebrates by peristaltic waves of muscular contraction. In another part of the tract movement is voluntary due to presence of skeletal muscles. The tongue is not mobile by itself. Many predatory fishes appear to regurgitate large food items from the stomach with great facility. This is possible because of pronounced development of striated muscles in the wall of the oesophagus to stomach region.

Intestinal Surfaces

The mouth cavity oesophagus and stomach are lined with soft mucus membrane as in the rest of the tract. There are no salivary glands except in parasitic lampreys. However, tract wall is liberally supplied with glands that secrete mucus, which lubricate passage of food material and protect the gut lining. The gut is highly elastic and permits larger size food masses. The linings of the small and large intestines are highly absorptive. The absorptive capacity of these areas is increased by throwing the walls into lengthwise folds (typhosole), transverse folds (rugae) and finger like projections (villi). Along the course of the tract, there are many gland cells that contribute digestive enzymes.

Glands and Digestive Enzymes

There are no digestive juices secreting into the oesophagus. The food passes very quickly from oesophagus to the stomach. Gastric glands occur in most of the predatory fishes. These glands secrete gastric juice, which contain HCl and pepsinogen, effective in combination to split large amount of protein molecules. In some carnivorous fishes, gastric acidity of pH 2.4 to 3.6 has been measured. Evidence for stomach enzymes other than peptidases are not clear. Some fishes (minnows) lack gastric gland and on this basis may not possess true stomach. Similarly, fish gizzard does not have digestive glands. HCl produced in the stomach facilitates in

1. Disinfecting action-killing bacteria.
2. Converts disaccharides to monosaccharides.
3. Activates pepsinogen to pepsin; pepsin is a proteolytic enzyme and hydrolyses the protein complex food into simple protein molecule in presence of HCl.

Pyloric caeca may have digestive or absorptive function, or both. An enzyme lactase and high levels of *saccharase* (invertase) has been found in pyloric caeca in trout and carp, which mainly feed on vegetable matter. Pyloric caeca and intestinal mucosa are sources of an enzyme lipase which breaks down fats into fatty acid and glycerol.

The liver is the largest gland of the body, which secretes biles, a product of both excretory and secretory activities of the organ. The bile is secreted by the hepatic cells into the bill capillaries and then it is collected into hepatic ducts, which join to form a common bile duct. Aystic duct connects it to the gall bladder. Gall bladder acts as a storehouse for continuously secreting bile.

In some fishes, only vestigial gall bladder is present and in others gall bladder is completely absent. The bile contains the fat emulsifying bile salts along with bile pigments, billierdin and bilirubin that originate from the breakdown of RBC and haemoglobin.

Besides its role in digestion, the liver also acts as storage organ for fats and carbohydrates; it has further important functions in blood cell destruction and blood chemistry as well as other metabolic functions such as production of urea and compounds concerned with nitrogen excretion. Liver also acts as a storage organ for fats and vitamin A and D. The content of vitamins in the liver of tunas is so high that persistent eating of their liver may lead to *hypervitaminosis.*

Pancreas is a glandular organ lying close to the duodenum; it is formed of exocrine and endocrine tissues. The exocrine tissue produces pancreatic juice which is carried by the pancreatic duct into the duodenum. It is neutral to alkaline and is important in digestion of food. Pancreatic juice contains enzymes for the digestion of proteins, carbohydrates and fats and nucleic acids. The pancreatic secretion is a complete digestive juice because it contains carbohydrate (CHO), fat and protein splitting enzymes.

The carbohydrate splitting enzymes from the pancreas is pancreatic *amylase* and it acts upon starch and glycogen in a similar but even more effective way than salivary amylase of mammals, completing the conversion of starch into maltose. The fat splitting enzyme pancreatic lipase, hydrolyzes each fat molecule into one molecule of glycerol and three of free fatty acids. The work of lipase is facilitated by the action of bile. There are no digestive enzymes in the bile. It breaks up large globules of fat into very small ones, giving much more surface for the enzyme to act upon on. The protein splitting enzymes of the pancreatic juice are *trypsin* and *chemotrypsin*. Trypsin is secreted in an inactive form, *trypsinogen*. This substance is converted into active trypsin through the action of another enzyme by the breakdown of large protein and polypeptide molecules into similar molecules. This process is by hydrolysis with two enzymes acting inside the molecules (endopeptidases) rather than at the ends. Following the action, the polypeptides are broken into much smaller units made up of 2,3,4 or more amino acids linked together. This is accomplished through the action of enzyme *carboxypeptidases*. This enzyme is also present in pancreatic juice.

The intestinal fluid contains, in addition to *enterokinase* mentioned above, several enzymes which are necessary to complete digestion of food into simple absorbable substances. The small intestine secretes a group of aminopeptidases and dipeptidases (Erepsin), which complete the breakdown of proteins into amino acids, each enzyme being quite specific as to which amino acid it will split off. The intestinal fluid also contains three inverting enzymes by which the disaccharides are split into monosaccharides. They are maltase-splitting maltose into glucose, lactase-splitting lactose to glucose and galactose and sucrose -splitting sucrose to glucose and fructose.

Absorption and Assimilation of Food

Absorption can be defined as "the passage of food through the lining of the digestive tract into the blood".

In order for digested food to be absorbed, they must be in aqueous solution; hence, they themselves must be soluble. The component molecules must further be of a size that will enable them to cross the membrane of the cells lining the tract, pass into the circulatory system and ultimately be carried to and enter the cells that need and store them. It is interesting to note in this context that fat absorption is intensified in the pyloric stomach of some fishes and pyloric caeca of others. Fats have been shown to enter into the lymph ducts in these regions without being split into their component fatty acid and glycerol molecules upon which depends the intestinal absorption depends. No absorption takes place in mouth. Not much absorption in the stomach also, except for simple molecules of glucose to some extent if at all they are present.

Mechanism of Absorption

Nearly all organic and inorganic compounds are absorbed in small intestine. Modifications of gut facilitate the process of absorption. The lining of the small and large intestine are highly absorptive. The absorptive capacity of these areas is increased by throwing the walls into length was rugae or villi (tryplosole). These folds are covered with epithelium within which is a net work of blood capillaries and also lymph vessels in absorption material passes from the lumen of the intestine through the epithelium and into the capillaries or lymph vessels by process known as active transport. The active transport involves movement of materials against concentration gradients and needs energy some substances. On the other hand it penetrate passively and diffuse through these folds.

Absorption of Proteins

The amino acids formed due to digestion of proteins in the intestine, do not accumulate there. But they are absorbed into the intestinal capillaries and these enter the portal veins, to be carried into the general circulation by the way of liver. Absorption is more rapid than simple diffusion. Some absorption of protein derivatives has been shown to occur; it has been shown that the smallest units into which the proteins are broken down in the intestine are dipeptides. The apparently leaves to intracellular digestion for the final breakdown of proteins into single building blocks (amino acids) from which the fish proteins are resynthesized. The amino acids absorbed into the blood diffuse through the body fluids and so reach all the tissue cells. At the same time, most of the tissue proteins are continually undergoing disintegration to release amino acids, which also enter the circulation and this become the "amino acid proff". From the pool amino acids are taken up by the cells to be built up into the cell structure as required, if the cell takes up as much amino acid as it losses. It is the state of dynamic equilibrium if the loss is greater, the cell wastes and the gain is greater the cell grows.

Ingested complex protein $\xrightarrow{\text{pepsin}}$ Simple amino acids ⟶ Blood stream or lymph vessels (to reach all tissues)

At the same time,

Tissue protein ⟶ amino acids ⟶ Blood streams (AA pool) ⟶ build up of cells

Absorption of Carbohydrate

This absorption of sugars from the stomach and colon is very slight. The proximal part of the small intestine seems to be the chief site for absorption of sugars. Simple diffusion can play some part no doubt in the absorption, when concentration in the gut exceeds that in blood, but the main features of absorption can only be explained by active transport, requiring energy. The absorption of sugars also depends upon the presence of Na^+ ions, probably also on Na : K ratio. However, there remains much to be found about the details of sugar absorption.

$$\text{Disaccharides} \xrightarrow{\text{saccharases (invertase)}} \text{monosaccharides}$$

Absorption of Fats

Fat absorption from the intestine is not clear. It is shown that long chained fatty acids are absorbed very largely into the lymph, whereas, short chained fatty acids enter the portal blood. It has also been pointed out by some that efficient fat absorption requires both lipase enzyme and bile salts.

$$\text{Long chained FA} \xrightarrow{\text{lipases, bile salt}} \text{simple fatty acids}$$

General Utilization Pattern of all Major Nutrients

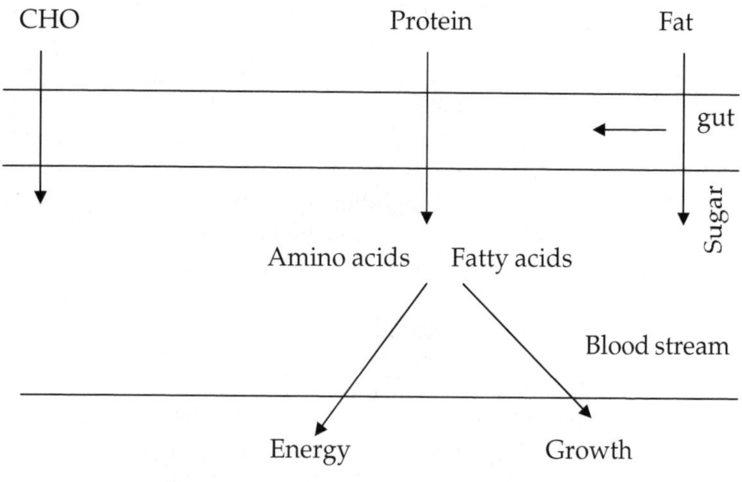

Physiology of Digestion in Shellfishes

In the aquatic environment, particularly in the sea, the crustacean has exploited every type of niche and this ecological diversity is paralleled by the diversity of eaten food. The micro crustaceans typically feed on microalgae, whereas that of larger crustaceans ranges from burrowing detritus feeders to active predators of molluscs and fish. Most of our knowledge of digestive physiology is derived from the larger Decapoda, although there has been appreciable research on the Amphipoda and Isopoda. Earlier knowledge of function was fragmentary, but over the last three decades experimental research on digestive physiology, using modern methods has grown considerably. However, there are still large gaps in our knowledge and much remains to be done.

The Principal Regions of the Gut and their General Functions

The Foregut: Structure and Function

Structure

The foregut or proventriculus is always divided into an anterior distendable part that usually serves as a crop in macrophage feeders. The posterior end of the chamber has a gastric mill, comprised principally of median dorsal and two lateral ossicles. The gastric mill leads into posterior part of the proventriculus, which in turn is divided into dorsal and ventral chambers. The dorsal chamber, which bears lateral grooves, leads into the midgut or directly into the hindgut. The ventral chamber contains W-shaped filter press, which leads into the digestive gland. The floor of the anterior proventriculus bears a median groove and two ventro lateral grooves, with fringing dense setae. The ventro lateral grooves lead to the filter press. There is a complex system of muscle attachment over the surface of the proventriculus particularly around the gastric mill.

Function

As the food enters the anterior chamber of the proventriculus, it is penetrated by this fluid from the digestive gland that flows towards dorso laterally in grooves in the posterior chamber. Trituration and further mixing with fluid occurs at the gastric mill ossicles. The food mass is continually being manipulated by the lateral plates of the anterior chamber and forced into the gastric mill. Eventually, the fluid passes from the food mass into the ventral grooves of the anterior chamber. Dense setae exclude larger particles and the fluid passes backward through the filter press which excludes particles above turn and finally into the openings of the digestive gland. Fluid from the digestive gland is pumped dorsally into the dorso lateral grooves, joined by the fluid squeezed from the food mass in the posterior chamber. Some fluid is also pumped in and out of the anterior diverticula of the midgut. The combined fluid then passes towards to the anterior chamber. The circulation is driven by the pumping action of filter press and associated structures. We do not know what happens to the fluid that enters digestive gland. Probably, dissolved nutrients are absorbed and the fluid with addition of more enzyme returns to proventricular circulation. This will be the trait area of research to be investigated.

Carnivorous	*Omnivorous/Herbivorous*
Intestine short	Intestine long
Stomach large and developed	Stomach poorly or under developed
Digestion of protein and fat efficient, except carbohydrate (CHO)	Digestion of CHO efficient, except protein and fat
Nutrient absorption more	Nutrient absorption less
Unable to absorb CHO due to insufficient enzymes	Able to absorb CHO due to sufficient enzymes
Digestive process fast	Digestive process slow
Less storage capacity of gut	High storage capacity of gut

Generally, Shellfishes (shorter intestine) normally fed continuously than finfishes (long intestine) because digestion is rapid and completed within 4–6 hrs.

Self Assessment Questions

1. Simply, what is digestion?
2. How protein is absorbed in fish body?
3. How carbohydrate (CHO) is absorbed in fish body?
4. How fat is absorbed in fish body?
5. Compare carnivorous Vs. omnivorous/herbivorous digestion.
6. How digestion is varied between fin and shellfishes?
7. Provide flow line for general utilization of all major nutrients in fish body?

8. List out the enzymes involved in protein, CHO and fat digestion in fish body.
9. Provide flow line for digestion in general in fish body.
10. Provide flow line for protein absorption in fish body.
11. Provide flow line for CHO absorption in fish body.
12. Provide flow line for fat absorption in fish body.
13. Why shellfishes fed continuously than finfishes?
14. Draw a diagram for digestive pattern for herbivorous fish.
15. Draw a diagram for digestive pattern for carnivorous fish.
16. Draw a diagram for digestive pattern for shrimp.

Chapter 4

Food and Feeding Habits of Fish

Diverse habitats of fish include 41.2 per cent in freshwater, 58.2 per cent in marine and 0.6 per cent an diadromous, migrating between fresh and saline waters. Habitats of fish shows wide range of feeding habits of fish include the following:

1. **Carnivores** (Predators): They have well developed grasping and holding teeth. Well defined stomach with strong acid secretions is present. Intestine is shorter than herbivores. Nocturnal fishes rely largely on smell, touch, and taste and also on their lateral line to locate and catch the prey (Sea bass, trout, eel, salmon, freshwater shark, catfish, murrel) and also possess well developed grasping and holding teeth.
2. **Herbivores** (Grass carp, silver carp and milk fish)
3. **Omnivores** (Common carp, shrimp)
4. **Detritivores** (Mud carp, mullet)
5. **Scavenger** (shrimp)
6. **Parasites**: It is perhaps the most highly evolved feeding habit. Parasites suck the body fluid from the host fish after making a hole in the side of the body. The males of some deep sea fishes are obligatory parasites on females of the

same species. Shortly after hatching the male finds a female and attach to the body. The female oblingly responds by developing a fleshy papilla from which the male fish can absorb nutrients, as it cannot take free living food.

7. **Grazers** (biting habits) – Parrot fishes – They graze the feed just like cows and sheep.
8. **Strainers** (plankton eating fish): Here, food objects are selected by size and not by kind. They can filter the water at the rate of 1–2 gallons/minute through gill rakers to retain plankton. In such fishes, filtering organs (Gill rakers) are well developed.
9. **Suckers** (presences of inferior mouth and suckling lips): They are mostly bottom feeding fishes. Inferior mouth and sucking lips are present. Fishes that suck in mud to extract organisms in it may not get gapped mouthful of food with each ingestion.
10. **Modification of lips, mouth shape, teeth, gill rakers and digestive tube**: Not all the fishes have stomach, *i.e.* a portion of digestive tube with a typical acid secretions and a destructive epithelial lining different from that of intestine. For example, in plant feeding roach fishes epithelial tissue of the oesophagus grades directly into that of the intestine. In other grazers such as parrot fishes analogous conditions are found. Some carnivores also have lost their stomach. The primary criterion for being able to do without the stomach does not seem to be whether fish is a herbivore of a carnivore but whether accessory adaptations for trituration and very fine grinding of food exist either in the form of teeth or a grinding apparatus such as gizzard. Where stomach exists, most pronouncedly in carnivores, they are characterized by a low pH and the prominent presence of pepsin among other digestive juices.

    The intestine too, has many variations; it is shortened in essential carnivores perhaps because meaty food can be digested more readily than vegetable ones. On the other hand it is often elongated and arranged in many folds predominantly in herbivores. In certain fishes the intestine itself seems to undergo digestion (*autolysis*) in fishes that cease feeding as sexual maturity and breeding arrive. Once the functional digestive tract becomes more thread with practically no lumen by the time the spawning is over and death approaches.

Stimuli for Feeding

There are two kinds of stimuli for feeding.

(*a*) The factors affecting the internal motivation or drive for feeding. These factors can be season, time of day, light intensity, temperature, time and nature of last feeding any internal rhythm.

(*b*) Food stimuli perceived by senses like smell, taste, sight and lateral line system. The interaction of these two factors determines the feeding of fish.

Self Assessment Questions

1. Is it necessary to study the food and feeding habits of fish before making aqua feed? If so, why?
2. Differentiate carnivores and herbivores.
3. Differentiate omnivores and herbivores.
4. What is copraphagus fish?
5. Name the mouth pattern of carps being cultured in composite fish culture practice.

Chapter 5

Nutritional Requirements of Cultivable Fin and Shellfishes – Proteins

Proteins are composed mostly of amino acids linked with peptide bonds and cross linked between chains with sulphydral and hydrogen bonds. Proteins are composed of carbon (50 per cent), nitrogen (16 per cent), oxygen (21.50 per cent) and hydrogen (6.50 per cent), sometimes phosphorus and sulphur. Proteins are considered as first abundant group of organic compounds (largest molecules in cell) in fish. It is also referred as one of the major nutrients also macromolecules. The empirical formula of amino acid is R-CH-NH_2–COOH. The gross energy of protein is 5.60 kg Cal/g. Because protein is the most expensive part of fish feed, it is important to accurately determine the protein requirements for each species and size of cultured fish.

Although over 200 amino acids occur in nature, only about 20 amino acids are common (About 20 major AA make up most proteins).

Of these, 10 are essential (indispensable) amino acids that cannot be synthesized by fish. The 10 essential amino acids that must be supplied by the diet are: methionine, arginine, threonine, tryptophan, histidine, isoleucine, lysine, leucine, valine and phenylalanine.

Of these, lysine and methionine are often the first limiting amino acids which are least abundant in proportion to its demands. Fish feeds prepared with plant (soybean meal) protein typically are low in methionine; therefore, extra methionine must be added to soybean meal based diets in order to promote optimal growth and health. It is important to know and match the protein requirements and the amino acid requirements of each fish species reared.

Fish are capable of using a high protein diet, but as much as 65 per cent of the protein may be lost to the environment. Most nitrogen is excreted as ammonia (NH_3) by the gills of fish, and only 10 per cent is lost as solid wastes. Accelerated eutrophication (nutrient enrichment) of surface waters due to excess nitrogen from fish farm effluents is a major water quality concern of fish farmers. Effective feeding and waste management practices are essential to protect downstream water quality.

- Protein generally is the most expensive component in fish diets.
- Carcass whole fish contains 75 per cent water, 16 per cent protein, 6 per cent lipid, 3 per cent ash.
- Gross energy of protein – 5.6 Kg cal./g.

Functions of Protein

- As a source of energy.
- Required for formation of hormones, enzymes.
- As a source of amino acid requirements for synthesis of diverse kind of proteins.
- To repair worn (or) wasted tissue and to rebuild new tissue.
- Serve as lubricants and protective agents.
- Serve as substrates for CHO and FA synthesis.

Classifications

Classifications are three major types:

1. **Fibrous protein** – Collagen, Elastin, Keratin.

2. **Globular protein**

Albumins	Globulins	Histones occur in cell nucleus with DNA
↓	↓	
Egg	Egg	
Milk	Milk	
Blood	Blood	

3. **Conjugated protein** – Glyco protein (mucous secretion), lipo protein (cell membrane), Metallo proteins, Phosphoproteins (milk, egg, york), Chromo proteins, Nucleo proteins, Cell nucleus

Amino Acid Classification Based on Chemical Structure

- ☆ Mono amino mono carboxylic acids (glycine, valine, threonine, leucine, iso leusine)
- ☆ Mono amino dicarboxylic acids (aspartic acid, glutamic acid)
- ☆ Diamino mono carboxylic amino acid (arginine, lysine)
- ☆ Sulphur containing acid (cystine, cysteine methionine)
- ☆ Aromatic and heterocyclic AA (phenyl alanine,tyrosine, typtophan, histidine, proline).

Types of Amino Acids

1. **Essential Amino Acids** are those which cannot biosynthesized de novo (again) by fish and hence to be provided in the diet. (methionine, arginine, threonine, tryptophan, histidine, isoleucine, lysine, leucine, valine and phenylalanine). These amino acids are required for biological processes.
2. **Non Essential Amino Acids** are those amino acids which can be synthesized de novo (again) by fish and hence need not be provided in the diet. (Proline, glutamic acid, aspartic, hydroxylproline). It dose not mean that these amino acids are not required by fish. These amino acids are not required for biological processes.
3. **Semi Essential Amino Acids** are those which cannot be biosynthesized de novo (again) by fish in sufficient level and supplied in lesser extent (Tyrosine, glycine cystine, serine).

Protein Requirement

Protein levels in aquaculture feeds generally average 18-20 per cent for marine shrimp, 28-32 per cent for catfish, 32-38 per cent for tilapia, 38-42 per cent for hybrid striped bass. Protein requirements usually are lower for herbivorous fish (plant eating) and omnivorous fish (plant-animal eaters) than they are for carnivorous (flesh-eating) fish, and are higher for fish reared in high density (recirculating aquaculture) than low density (pond aquaculture) systems.

Protein requirements generally are higher for smaller fish. As fish grow larger, their protein requirements usually decrease. Protein requirements also vary with rearing environment, water temperature and water quality as well as the genetic composition and feeding rates of the fish. Protein is used for fish growth if adequate levels of fats and carbohydrates are present in the diet. If not, protein may be used for energy and life support rather than growth.

Carnivorous fish need 40-50 per cent; Omnivores fish need 25-35 per cent.

Warm period and tropical climate – require lesser protein and carbon and vis-a-versa.

Protein Requirement of Selected Fish Species

Common carp	330-380; 250-350 g/kg feed
Grass carp	360-420; 300 g/kg feed
Rohu	400-450; 300 g/kg feed
Catla	470; 300 g/kg feed
Mrigal	400-450 g/kg feed
Silver carp	370-420; 300 g/kg feed
Channel cat fish	520; 320-360 g/kg feed
Clarias sp.	357 g/kg feed
Murrel	520 g/kg feed
Tilapia	400; 280-300 g/kg feed
Milk fish	350-400 g/kg feed
Mullet	350- 400 g/kg feed
Grouper	400 g/kg feed
Seabass	387 g/kg feed
Prawn	40 per cent

Shrimp	22-60 per cent
P. indicus	43 per cent
P. monodon	35-46 per cent
P. azetecus	as low as 22 per cent
P. japonicus	as high as 60 per cent
Rainbow trout	40-45 per cent
Atlantic salmon	45 per cent

Amino Acid Requirements for Fish and Shellfish

Amino Acid	*Fish (g/100 g protein)*	*Shrimp/Prawn (g/100 g protein)*
Methionine	2.6	0.86–1.08
Arginine	3.6	2.09–2.61
Threonine	2.4	1.30–1.62
Tryptophan	0.50–1.50	0.29–0.36
Valine	2.4	1.44–1.80
Isoleucine	2.4	1.26–1.58
Leucine	3.5–5.3	1.94–2.43
Phenylalanine	5–6.5	1.44–1.80
Histidine	1.5–2.1	0.76–0.95
Lysine	3.6	1.91–2.39

Factors Affecting Protein Requirement

- ✰ Size and age
- ✰ Fertility of the culture system
- ✰ Levels of management and intensification
- ✰ Seasons
- ✰ Geographic location.

Protein Sources – Predominantly Used

Animal

Fish meal, squid meal, clam meal, mussel meal, crab head meal, prawn head meal, squilla meal, silkworm pupae, poultry waste meal, slaughterhouse waste.

Plant

Soybean meal, wheat products blood meal, yeast, and cotton seed meal, peanut meal, corn glutens meal, rice bran, wheat bran, ground nut oil cake.

Protein Estimation

1. Kjeldahl method → higher protein
2. Biuret method
3. Folin-lowry's method

Limiting Amino Acids

The EAA content of different feed ingredients varies even more widely. This is one of the principal reasons why a compound diet, made from several ingredients, is potentially more efficient than a single ingredient, which may be too high or too low in one or more essential amino acids. The amino acid profile of a feed must be balanced for the dietary protein to be used effectively. This can be illustrated in the following way. Suppose the exit gate of a pond is composed of ten vertical planks of wood (Figure 1), each of which represents one of the essential amino acids, numbered one to ten.

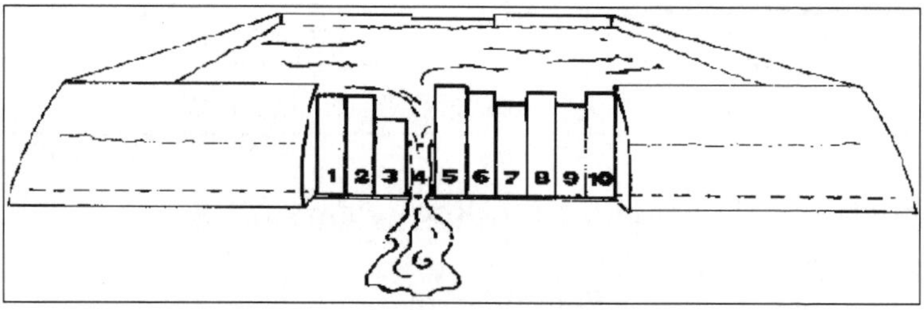

Figure 1: Limiting Amino Acids (A)

It will be seen from the diagram that the level of water in the pond will depend on the height of the shortest plank (plank 4). The shortest plank represents the 'first limiting amino acid'. If this plank is lengthened (or this amino acid is increased in level by altering the proportion or the type of ingredients used, or by adding it in synthetic form) then plank 3 would control the water level (or be the next limiting amino acid). Ideally all the planks should be just as high as the level of water desired in the pond (the quantity of each amino acid in the diet at exactly the correct level for the species being cultured) to avoid wastage of materials. Figure 2 illustrates what happens if one plank is unnecessarily long (or if one amino acid is present in excess in the diet)–it serves no useful purpose and is an unnecessary expense.

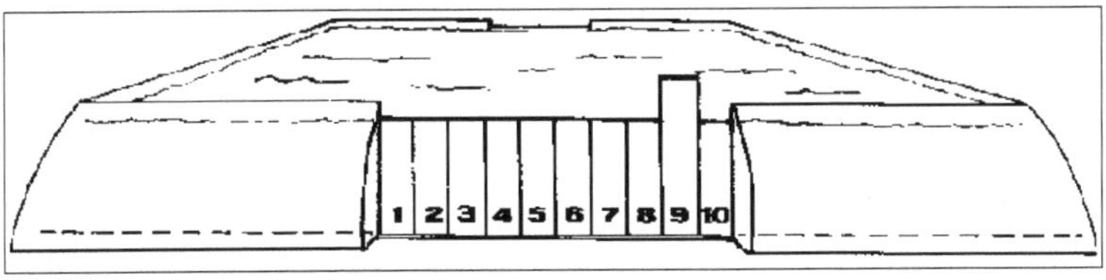

Figure 2: Limiting Amino Acids (B)

An unbalanced diet, particularly where one or more EAA's are deficient, is one of the principal reasons why the dietary protein level may have to be unnecessarily high to promote optimum growth. In such a case the rest of the protein is wasted in order to supply the required level of the deficient EAA(s).

Even when an EAA is shown by chemical analysis to be present in sufficient quantities it may not be biologically available to the animal. For example, the free a -amino group of lysine may become bound to other molecules during processing of the feedstuff, rendering it unavailable to the target animal. Thus the method of processing and the quality of high protein (expensive) ingredients is an important factor in formulation of compound feeds.

Methionine and phenylalanine are amino acids which can be 'spared' or partially replaced by cystine and tyrosine (two non-essential amino acids) in the diet. Up to approximately 50 per cent of either the methionine or the phenylalanine requirement can be spared or replaced by cystine or tyrosine respectively. Methionine and lysine are usually the first two limiting amino acids in feeds. In tables of the levels of EAA's in ingredients a combined figure is often given for methionine and cystine. While fish, and particularly shrimp, do not thrive well on mixtures of pure synthetic amino acids instead of protein, individual amino acids (notably 1-lysine and dl-methionine) can effectively be used to supplement diets otherwise deficient in these amino acids. Whether synthetic amino acids are used or not, as usual, depends on the economics of the situation.

The amino acid profile of examples of some high-protein sources are compared with that of chicken egg protein (considered an excellent source of protein for animals). It will be seen that most plant proteins are deficient in the sulphur-amino acids (methionine and cystine) but that the 'score' of sesame and palm kernel cake is higher than fish meal in this respect. Mung and red beans (legumes) are high in lysine. Meat meal, because of its poor levels of iso-leucine and methionine and cystine, is a poor quality protein compared with fish meal. Only milk does not have seriously limiting levels of methionine and cystine, compared with fish meal. Fish meal is acknowledged to be the best (but usually the most expensive) protein source for fish feeds. Feeds should, ideally, be matched (balanced) with the specific EAA requirements of the species being cultured but this information is, as yet, incomplete for most aquatic species. 'Non-essential' amino acids have an important role to play in diet palatability.

The quantitative Essential Amino Acid (EAA) requirements of different species varies.

- ☆ EAA requirement may vary within and between species.
- ☆ EAA requirement may vary in feed ingredients.

  AA profile of a feed must be balanced for the dietary protein.

Methods of Determination for Essential Amino Acid (EAA) Requirement

Direct Methods

In this, as one at a time basis, amino acid is deleted from the amino acid test diet and dose- response growth curve is made. Dietary requirement is taken at break point.

Indirect Methods

Five methods are available, they are:

Nitrogen Balance Method

This is modified method of direct method. Quantitative variation in free amino acids levels in specific tissue pools such as whole blood, plasma, haemolymph, or muscle is made with reference to the deleted amino acids.

Limiting Amino Acids and Chemical Score of Essential Amino Acid Content of Selected Feed Proteins[#]

Feedstuff	*Arg*	*His*	*Iso*	*Leu*	*Lys*	*Met + Cys*	*Phe*	*Thr*	*Try*	*Val*
Fish meal (Peru)	85	85	66	88	110	71	78	74	58	61
Meat meal	77	96	28*	100	86	36*	72	60	68	75
Milk, skimmed	53*	92	88	110	104	69	91	80	73	75
Milk, whole	60	100	100	136	106	83	92	83	84	78
Groundnut oil cake	164	92	43*	72	53*	24*	91	51*	- *	45*
Coconut oil cake	164	78	43*	71	37*	34*	76	54*	- *	57
Soybean meal	110	89	66	92	90	54*	102	69	68	63
Palm kernel cake	207	92	54*	75	54*	83	66	64	147	69
Cottonseed cake	164	96	46*	69	60	51*	100	58	58	55
Sunflower oil cake	112	59	46*	62	32*	22*	61	47*	79	46*
Sesame oil cake	191	107	51*	88	42*	94	79	58	73	60
Mung bean	100	78	66	84	107	31*	109	62	-	62
Red bean	112	130	77	49*	128	27*	107	83	42*	72
Chlorella vulgaris	77	55	64	90	44*	47*	92	100	68	72
Spirulina maxima	97	66	86	94	67	33*	92	83	73	79
Scenedesmus obliquus	83	55	63	109	84	40*	85	94	73	88
Torula yeast	77	81	68	94	111	51*	81	91	63	66
Brewer's spent grains	68	66	77	97	48*	22*	87	58	68	66

Source: ADCP, 1985.

* Seriously limiting amino acids.

\# Scores based on comparison with whole egg protein of the following amino acid composition (percentage of protein): arginine, 6.7; cystine, 2.2; histidine, 2.7; isoleucine, 7.0; leucine, 8.5; lysine, 6.8; methionine, 3.3; phenylalanine, 5.4; threonine, 5.5; tryptophan, 1.9; and valine, 8.2.

Tissue Culture Method

Like dietary deletion, specific amino acids free media are used.

Administration of Radio Isotopic Assay

Animal is fed or injected with one of the radio activity labeled readily metabolizable metabolites such as (^{14}C) glucose, $^{14}CO_2$, (^{14}C) acetate, (^{14}C) succinate, (^{14}C) pyruvate. Fish or part of the tissue is latter assayed after a period of incubation. Non essential amino acids are being able to be synthesized from a precursors take up labeling, while, essential amino acids remain unlabelled.

Starve and Feeding Method

In alternatively starved and fed animals, fluctuation in free amino acids levels are made in tissue pools; where in essential amino acids fluctuate drastically between feeding and starvation, while, non essential amino acids remains steady.

Ogino's Carcass Deposition Method

This is the only method devised to determine quantitative requirement for essential amino acids, specifically for fishes. Ogino's observed similarity in percentage composition existing between dietary essential amino acids requirements of fishes and essential amino acids profile of fish muscle. His procedure is to estimate daily nitrogen or protein rate, per cent feeding rate for 100 g body weight, per cent digestibility for protein and for each amino acids further test animals. From these parameters, he calculated optimum level for amino acids required to be present in the dietary protein source and optimum dietary requirement/day for each amino acid.

Self Assessment Questions

1. What are proteins?
2. What is the calorific value of proteins?
3. List out the different functions of proteins.
4. Classify generally the proteins.
5. Classify the different types of amino acids.
6. Differentiate between Essential and Non-essential Amino Acids.
7. Expand EAA, NEAA and SEAA.

8. Collect the protein requirements for locally commercially available fishes.
9. What are the different factors affecting protein requirements?
10. What are the protein sources (both plant and animals) used for fish feed preparation?
11. What are the methods available for protein estimation?
12. Define limiting amino acids and give examples.
13. What are the methods of determinations for essential amino acid requirements?

Chapter 6

Nutritional Requirements of Cultivable Fin and Shellfishes – Fats (Lipids)

Next to proteins, lipids represent the second most abundant group of organic compounds in the animal body. In feedstuff chemistry, the words fat, lipid and oil are sometimes used synonymously. Tables of feed composition often refer to the crude fat level, by which is meant the material which can be removed from the feed by ether extraction. The term 'oil content' is also often used in this context. The term 'crude' lipid content can also be used. The word lipid is a general term which covers sterols, waxes, fats, fat soluble vitamins, phospholipids and sphingomyelins.

Many of the vitamins are fat soluble and will be extracted by ether–thus the term crude lipid content. The words oil, fat, and wax, reflect the increasing melting points of these lipid components.

Lipids (fats) are high energy nutrients that can be utilized to partially spare (substitute) protein in aquaculture feeds. Lipids supply about twice the energy as proteins and carbohydrates (Lipids exhibit high calorific value (8-9.5 Kcal/g) than proteins and CHO).

Dietary lipids are a concentrated and highly digestible source of energy and are a source of essential fatty acids (EFA) which are necessary for normal growth and survival of all animals. Lipids, typically comprise about 15 per cent of fish diets, supply essential fatty acids (EFA) and serve as transporters for fat soluble vitamins.

Fats are the fatty acid esters of glycerol and are the primary means by which animals store energy. Fish are able to metabolize lipids readily particularly when deprived of food, as during the migration of salmon, for example. Phospholipids are components of cellular membranes. Sphingomyelins are found in brain and nerve tissue compounds. Sterols are important components or precursors of sex and other hormones in fish and shrimp. Waxes form important energy storage compounds in plants and in some animal components.

General Function of Fats

- ☆ Serve as a biological carriers for the absorption of the fat soluble Vitamins A, D, E and K.
- ☆ As a source of essential fatty acids (EFA).
- ☆ As a source of essential steroids which play an important role in biological function.
- ☆ Act as lubricants for the passage of feed through pellet diets which reduce the dustiness of feeds.
- ☆ Play a role in feed palatability.

Classifications

Two major types namely:

1. Glycerol based lipids and non glycerol based lipids (waxes, steroids, tarpenes).
2. Glycerol based lipids are of two types namely,
   (*a*) Simple (Fats and oils) and
   (*b*) Compound (two types–*Glycerolipids* – glycolipids, galactolipids and *Phospholipids*–lecithins, cephalins).

Fatty Acids

Fatty acids (FA) are components of lipids. Over 40 different FA are known to occur in nature. Fatty acids are represented by a general formula, $CH_3 - (CH_3)_n - OOH$.

Classification of Fatty Acids

Fatty acids are of two major types, namely, Saturated fatty acids–Butyric acid, Lauric acid, Caproic acid, Capric acid (Fatty acids without any double bond) and Unsaturated fatty acids–Palmitoleic acid, Oleic acids, Linoleic acid, Linolenic acid, Arachidonic acid, Eicosapentaenoic acid (EPA), Docosahexa enoic acid–DHA (Fatty acids with double bond). Saturated and unsaturated fatty acids are otherwise called non essential fatty acids and essential fatty acids respectively.

Unsaturated fatty acids are of three major types, namely, Mono Unsaturated Fatty Acid – MUFA (Oleic acid and Palmit oleic acid), Poly Unsaturated Fatty Acid–PUFA (n-6 series fatty acids such as Archidonic and Linoleic acid) and Highly Unsaturated Fatty Acid–HUFA (n-3 series fatty acids such as Linolenic, EPA and DHA).

The fatty acids, which are components of lipids, are categorized in the following way. They are given a common name but are also, besides their straightforward chemical formula, given a specific numerical designation, such as 14:0; 20:1; 18:3n-3; 18:2n-6; 20:4n-6; 22:5n-6; or 22:6n-3. This nomenclature refers to the length of the carbon chain in the molecule, the number of carbon-carbon double bonds present and the position of the first double bond. This can be illustrated by the chemical formulae of some of the fatty acids mentioned above. This nomenclature may sound complex to those less exposed to biochemistry but it is necessary just to know what the terminology means so that it is possible to understand references to different types of fatty acids when fish and shrimp nutrition is being discussed. The methyl group is shown in the following formulae as CH_3.

Linolenic acid: 18:3n-3

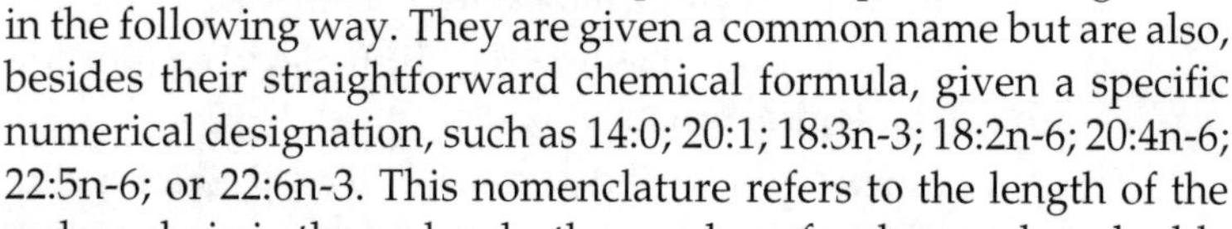

Linoleic acid: 18:2n-6

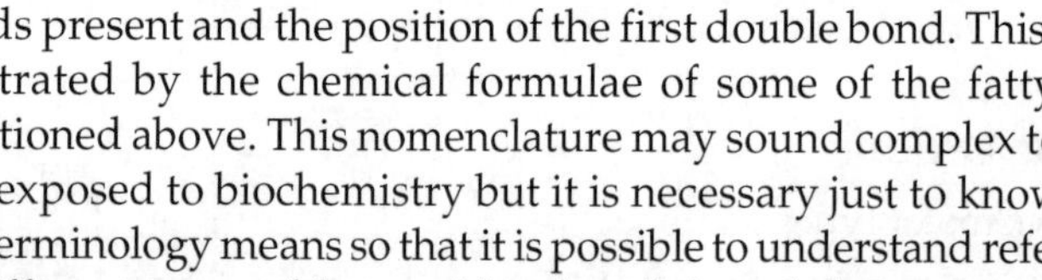

Arachidonic acid: 20:4n-6

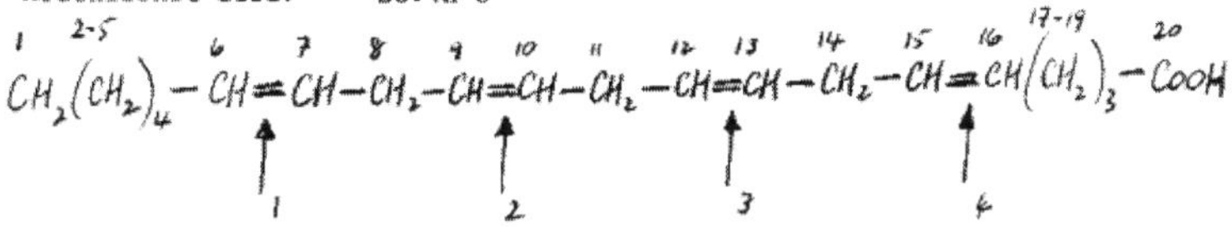

In the designation 20:n-6, for example, '20' means that there are 20 carbon atoms in the chain. '4' means that there are four carbon-carbon double bonds (the double bond is shown as a = sign and the carbon atom is shown as C). n-6 means that the first double bond, numbering from the methyl (CH_3) end, occurs after the sixth carbon atom in the chain. Myristic acid (14:0) has fourteen carbon atoms but no C=C double bonds. Arachidic acid (20:0) also has no double bonds. Linolenic acid (18:3n-3) has eighteen carbon atoms and three double bonds, the first of which appears on the third carbon atom. Linoleic acid (18:2n-6) has two double bonds and the first occurs on the sixth carbon atom and so on.

Those fatty acids which have their first double bond on the third carbon atom are known as the 'n-3' series or the 'linolenic' series after the name of the fatty acid in the series with 18 carbon atoms in its chain. Similarly, those which have their first double bond on the sixth carbon atom are known as the 'n-6' series or the 'linoleic series. It is essential to notice the difference (which is only the addition of the letter N) between the two words LINOLENIC and LINOLEIC.

Saturated fatty acids are those without any double bonds. Monosaturated fatty acids are those with only one double bond, while those with more than one double bond are known as polyunsaturated fatty acids. Those with fewer double bonds are referred to as 'more saturated' than those with a greater number. The n-3 series and n-6 series fatty acids, and the n-7 and n-9 fatty acids are all members of the group known as polyunsaturated fatty acids, because they have more than one double bond. Sometimes, these are abbreviated as PUFA's. Members of this group which have many (4 or more) double bonds are sometimes referred to as highly unsaturated fatty acids (HUFAs).

It is hoped that the above explanation will help extension workers and students to understand these terms, which are frequently referred in papers and books about fish and shrimp nutrition owing to their importance.

Simple lipids include fatty acids and triacylglycerols. Fish typically require fatty acids of the omega 3 and 6 (n-3 and n-6) families. Fatty acids can be: (*a*) saturated fatty acids (SFA, no double bonds), (*b*) polyunsaturated fatty acids (PUFA, >2 double bonds), or (*c*) highly unsaturated fatty acids (HUFA; > 4 double bonds). Marine fish oils are naturally high (>30 per cent) in omega 3 HUFA and are excellent sources of lipids for the manufacture of fish diets. Lipids from these marine oils also can have beneficial effects on human cardiovascular health.

Marine fish typically require n-3 HUFA for optimal growth and health, usually in quantities ranging from 0.5-2.0 per cent of dry diet. The two major EFA of this group are eicosa pentaenoic acid (EPA: 20:5n-3) and docosa hexaenoic acid (DHA:22:6n-3). Freshwater fish do not require the long chain HUFA, but often require an 18 carbon n-3 fatty acid, linolenic acid (18:3-n-3), in quantities ranging from 0.5 to 1.5 per cent of dry diet. This fatty acid cannot be produced by freshwater fish and must be supplied in the diet. Many freshwater fish can take this fatty acid and through enzyme systems elongate (add carbon atoms) to the hydrocarbon chain and then further desaturate (add double bonds) to this longer hydrocarbon chain. Through these enzyme systems, freshwater fish can manufacture the longer chain n-3 HUFA, EPA and DHA, which are necessary for other metabolic functions and as cellular membrane components. Marine fish typically do not possess these elongation and desaturation enzyme systems and require long chain n-3 HUFA in their diets. Other fish species, such as tilapia, require fatty acids of the n-6 family, while still others, such as carp or eels, require a combination of n-3 and n-6 fatty acids.

The necessity of high dietary levels of PUFAs in aquatic animal diets makes the possibility of fats becoming rancid very real. These may be toxic or depress growth. Some important points to remember in this context are:

1. Aquatic animals have a higher requirement for the n-3 series of fatty acids than terrestrial animals, for which the n-6 series is more important;
2. EFA deficiencies are more noticeable in seawater than in freshwater conditions (for trout). Thus, salinity affects EFA requirements;

3. Marine fish appear to have a greater requirement for HUFAs than freshwater or anadromous species. It is not yet known whether they can utilize the n-6 series as well as they can the n-3 series;
4. Coldwater species appear to have a greater requirement for the n-3 series fatty acids than warm water species;
5. Shrimps and prawns have a requirement for the n-3 series and the n-3:n-6 ratio is important;
6. The levels of either type of PUFAs can be detrimentally high in a feed. Knowledge of the specific requirements of a species is therefore constantly being sought to optimize formulation practice;
7. Although many vegetable lipids (but not those of palm, olive or coconut) are high in PUFAs, the best sources (and the most expensive) sources of the n-3 HUFAs are marine lipids. Vegetable oils tend to have high levels of the n-6 series (linoleic series). Lard and beef tallow have low total levels of PUFAs.

Dietary Lipid Requirements

Common carp	8-18 per cent
Indian and Chinese carps	5-8 per cent
Tilapia	6-10 per cent
Channel cat fish	8-12 per cent
Cat fish	10 per cent
Milk fish and Mullets	6 per cent
Seabass	12-13 per cent
Grouper	14 per cent
Eel	12-15 per cent
Rainbow trout fry	15 per cent
Fingerling	12 per cent
Adult	9 per cent
Shrimp – general	5-10 per cent
P. indicus	6-10 per cent

P. monodon	8-10 per cent
P. japonicus	8-10 per cent
Scampi	3-6 per cent

Essential Fatty Acids Requirements

Shrimp	
Linoleic acid	0.4 per cent
Linolenic	0.3 per cent
EPA	0.4 per cent
DHA	0.4 per cent
Common carp	0.5-1 per cent
Eel	0.5 per cent
Tilapia	1 per cent
Milk fish, mullets, sea bass, grouper and sea bream	1.2 per cent

Phospholipids

It represents the second largest lipid component next to fats and oil. Also as esters of fatty acids and glycerol in which positions 1 and 2 are esterified with fatty acids and position 3 with phosphoric acid and nitrogenous base. According to nitrogenous base, it is divided into lecithins and cephalins. Recommended level is 2 per cent. If lecithin source used, the dose is 1 per cent. Marine invertebrate oils contain high phospholipids. Squid, shrimp, clam oils contain 35-50 per cent–phospholipids. Soybean lecithin is often used as cost effective source of lecithin.

Dietary phospholipids have been shown to enhance the growth rate of crustaceans, but supplemental lipids such as lecithin are probably only necessary in purified diets (feeds made from purified ingredients, rather than the type of ingredient normally used in commercial animal feeds).

Sterols

Many sterols and essential components such as molting hormones, sex hormones, bile acids, vitamin D are synthesized from cholesterol. It also functions as a component of membranes and in the absorption and transport of fatty acids. Recommended level is 0.2–0.5 per cent Marine invertebrate meals and oils (clam, mussel, crabs, shrimp and squid) are good source of cholesterol. It is only essential for shrimps and not essential for diet of fish. Rich source of sterol is obtained from prawn head meal.

Self Assessment Questions

1. What are lipids?
2. What is the calorific value of lipid?
3. List out the different functions of fat.
4. Classify generally the lipids.
5. Differentiate between MUFA, PUFA and HUFA.
6. Differentiate between n-3/n-6 series fatty acids.
7. What are DHA and EPA?
8. Differentiate between saturated and unsaturated fatty acids.
9. Dietary lipid requirement of different cultivable fishes.
10. What are phospholipids?
11. What are sterols?

Chapter 7

Nutritional Requirements of Cultivable Fin and Shellfishes – Carbohydrates (CHO)

After the proteins and lipids, the carbohydrates (CHO) represent the third most abundant group of organic compounds in the fish body. CHO are the most abundant and relatively least expensive source of energy (cheap natural source of energy) in aquaculture. CHO are usually defined as substances containing Carbon, Hydrogen and Oxygen with the last two elements being present in the same ratio as in water *i.e.* $[C_x(H_2O)_y]$. CHO is stored in the form of glycogen and cellulose in animals and plants respectively.

Carbohydrates (starches and sugars) are the most economical and inexpensive sources of energy for fish diets. Although not essential, carbohydrates are included in aquaculture diets to reduce feed costs and for their binding activity during feed manufacturing. Dietary starches are useful in the extrusion manufacture of floating feeds. Cooking starch during the extrusion process makes it more biologically available to fish. In fish, carbohydrates are stored as glycogen that can be mobilized to satisfy energy demands. They are a major energy source for mammals, but are not used efficiently by fish. For example, mammals can extract about 4 kcal of energy from 1 gram of carbohydrate, whereas fish can only extract about 1.6 kcal from the same amount of carbohydrate. Up to about 20 per cent of dietary carbohydrates can only be used by fish.

The carbohydrates, which include starches, sugars, cellulose and gums containing only the elements carbon, hydrogen and oxygen are usually the cheapest source of energy in foods and feeds. Fish and shrimp, however, vary in their ability to digest carbohydrate effectively. Many fish appear to be able to utilize simple carbohydrates, such as sugars, more effectively than complex starches; the reverse appears to be true for shrimps and prawns, but this observation may be confused by the beneficial effect that carbohydrates tend to have on the structural integrity of the feed, caused by the binding quality of starches. Carnivorous fish such as salmon and trout and particularly, marine fish are not efficient converters of carbohydrate. Channel catfish, like shrimp, appear to be able to utilize complex carbohydrates more readily than simple sugars. Channel catfish and carp can utilize quite high levels of dietary carbohydrate; the natural diet of grass carp is very high in this component.

Some carbohydrates are normally regarded as indigestible. These are reported separately in the tables of feed composition as 'fibre' or 'crude fibre'. Fibre includes substances such as celluloses (from plants), lignin, chitin etc. Many fish do not have the enzyme cellulase which is necessary for the digestion of cellulose and fibre is usually regarded as unavailable as an energy source. At small levels, however, it may aid pelletability. Cellulase, however, is produced by the gut bacteria of many fish, as it is chitinase in crustaceans and herbivorous fish are able to digest fibre.

Function of Carbohydrates (CHO)

- ☆ As a cheapest source of energy.
- ☆ Aids in binding.
- ☆ Serve as precursors of various metabolic intermediates like non essential amino acids, nucleic acids and chitin.
- ☆ It increases feed palatability.
- ☆ Reduce the dust content of finished feeds.

Classification of Carbohydrates (CHO–According to the Chemical Structure)

The basic units of CHO's are known as monosaccharides. CHO

are of two types, namely, Sugars (those CHO which contain <10 monosaccharides units) has sweet taste; Non sugars (are those CHO which contain >10 monosaccharides units) and do not possess a sweet taste. Sugars are of following types:

1. Monosaccharides (glucose, fructose, galactose) give poor results because of quick and immediate absorption.
2. Disaccharides (sucrose, maltose, lactose) give good results (better utilization and growth) because of steady digestion and slow and gradual absorption.
3. Oligosaccharides–Tri saccharides (present in pulses like; Soybean and act as anti nutritional factors-raffinose); Tetra saccharides (starchy base).
4. Polysaccharides (dextrin, starch).

Non sugars are of 2 types; Homopolysaccharides (Chitin and cellulose) and Heteropolysaccharides

1. Homo polysaccharides (*e.g.* Starch, Dextrins, Glycogen, Cellulose).
2. Hetero polysaccharides – Hemicelluloses, Gums, Mucilages (from seaweeds likes *Gelidium* sp. *Laminaria* sp.)- Pectin substances, Muco polysaccharides.

Freshwater and warm water (higher intestinal amylase activity) species are generally able to utilize higher levels of dietary CHO than coldwater and marine species.

In the absence of dietary CHO, animals (shrimps) will utilize protein to meet their energy need. This relationship between protein and CHO has been referred to as the protein – sparing action of CHO.

Utilization of Dietary CHO

It is well illustrated by experiments known as glucose tolerance test. A large amount of glucose is administered orally to the fish. The blood concentration of glucose is measured over a time. In human beings, glucose in blood returns to normal about 1-2 hours after feeding. Fish respond to these much more slowly *i.e.* glucose in blood returns to normal at least 7 hour after feeding. It is apparent that CHO in fish diets has been carefully controlled since excess deposited as glycogen subsequently less readily available to fish for use as energy.

CHO Utilization in Fish

Carnivorous fish species (Eel, seabass, trout, sea bream) have poor ability to digest CHO and formulated feeds of these species contain CHO level less than 20 per cent. Omnivorous and herbivorous species are capable of utilizing 40-45 per cent CHO in the form of gelatinized starch. The traditional feed mix. GNOC and RB contain as high as 45 per cent CHO. Poor utilization of starch in carnivorous fish is due to low amount of amylase produced.

CHO Utilization in Prawns

In shrimp and prawn, CHO is an important in energy production, Chitin synthesis and NEFA synthesis. Chitin (0.5 per cent) or its precursor glucosamine (0.8 per cent) when induced in the diet has been shown to improve growth and feed efficiency in prawns and shrimps. In shrimp and prawn, Chitin is required in the form of exoskeleton as well as peritrophic membranes. CHO level in semi intensive culture system needs 25–30 per cent; CHO level in extensive culture system needs 35–40 per cent.

Self Assessment Questions

1. What is CHO?
2. Mention the unit of CHO.
3. CHO are composed of what?
4. Mention the different functions of CHO.
5. Classify the CHO.
6. Which fraction of CHO's is used in fish diet and Why?
7. Collect the CHO requirements for fish/shrimp.
8. State the CHO utilization in fish.
9. What are the four reaction pathways exist in CHO metabolism?
10. What is the gross energy value of CHO?

Chapter 8

Nutritional Requirements of Cultivable Fin and Shellfishes – Vitamins and Minerals

Vitamins

Vitamins are distinct from the major food nutrient *i.e.* Proteins, Lipids and CHO in that they are not chemically related to one another and are present in very small quantities within animal and plant food stuffs and are required by the animal body in trace amounts.

Vitamins are a diverse group of organic compounds necessary in the fish diet in minute quantities for normal growth, reproduction, health and general metabolism. They often are not synthesized by fish and must be supplied in the diet. Factors which affect requirement are size, age, growth rate, environmental conditions and nutrient relationships.

Classification of Vitamins

Water soluble vitamins (11 Nos.)–Vitamin B_1(Thiamine), Vitamin B_2 (Riboflavin), Vitamin B_6 (Pyridoxine), Vitamin B_{12} (Cynacobalamine), Vitamin C (Ascorbic acid), Pantothenic acid, Niacin, Biotin, Inositol, Choline and Folic acid. Of these, Vitamin C probably is the most important because it is a powerful antioxidant and helps the immune system in fish.

Fat soluble vitamins (4 Numbers): The fat-soluble vitamins include A vitamins, the retinols (responsible for vision); D vitamins, cholecaeciferols (bone integrity); E vitamins, the tocopherols (antioxidants); and K vitamins, the menadione (blood clotting, skin integrity). Of these, vitamin E receives the most attention for its important role as an antioxidant.

The most common vitamin deficiency in fish nutrition is that of vitamin B (thiamine). Moist or wet feeds containing raw aquatic animal products, especially if not fed immediately after manufacture; contain enzymes called thiaminases which may partially or completely inactivate the thiamine present in the feed. Thiaminase levels in freshwater fish flesh are higher than in that of marine fish (*i.e.*, the source of the fish flesh is important). Thiaminases have also been reported in other ingredients, such as rice polishing and beans. Supplemental thiamine may therefore be required in diets containing fish flesh unless they have been pasteurized.

Water Soluble Vitamins

Vitamin B_1

Role in controlling CHO metabolism. Thiamine is readily destroyed in the presence of mineral (Cu). Deficiency symptoms– Anorexia, poor growth, pigmentation and mortality. *Sources* – dried distillers solubles, fish solubles, rice bran, wheat middlings and yeast. Levels in feed: Carp – 2-3 mg/Kg; Shrimp – 50 mg/Kg; Salmon and trout–10 mg/Kg

Vitamin B_2 (Riboflavin)

It incurs high loss in feed processing. Synthesized by all plants and microorganisms; not by animals. Deficiency symptoms – Anorexia, poor growth, abnormal swimming behavior and mortality. *Sources*: dried distillers solubles, fish meal, fish solubles, liver meat and yeast. Levels in feed: Carp – 4-7 mg/Kg; Shrimp–30- 50 mg/Kg; Salmon and trout–30-50 mg/Kg.

Vitamin B_6 (Pyridoxine)

Required in transamination, deamination and decarboxylation. Deficiency symptoms – Poor growth, abnormal swimming behaviour and mortality. *Sources* – dried distillers solubles, fish meal, fish solubles, liver meat and yeast. Levels in feed: Carp – 5-10 mg/Kg; Shrimp–50 mg/Kg; Salmon and trout–10-20 mg/Kg.

Pantothenic Acid

Deficiency symptoms – abnormal gill features anorexia and mortality. *Sources* – dried distillers solubles, cotton seed meal, fish solubles, fish meal, pea nut meal rice bran wheat bran and yeast. Levels in feed: Carp – 30-40 mg/Kg; Shrimp–75 mg/Kg; Salmon and trout–60 mg/Kg.

Niacin

Deficiency symptoms – Anorexia, poor growth, lethargy and mortality. *Sources* –blood meal dried distillers solubles, cotton seed meal, fish solubles, fish meal pea nut meal, rice bran, wheat bran, yeast and corn gluten meal. Levels in feed: Carp –28 mg/Kg; Shrimp–200 mg/Kg; Salmon and trout -50 mg/Kg.

Biotin

Deficiency symptoms – Anorexia, slow growth and pigmentation. *Sources* – dried distillers solubles, cotton seed meal, rice polishings and yeast. Levels in feed: Carp – 0.6-1 mg/Kg; Shrimp – 1 mg/Kg; Salmon and trout–0.4 mg/Kg.

Inositol

Deficiency symptoms – Anorexia and slow growth. *Sources* – Fish meal, liver meal, wheat germ, soy lecithin and yeast. Levels in feed: Carp – 440 mg/Kg; Shrimp–300 mg/Kg; Salmon and trout–200-400 mg/Kg.

Choline

Deficiency symptoms – Anorexia, poor growth and fatty tissues. *Sources* – Fish meal, cotton seed meal, fish solubles, fish solubles shrimp meal, soybean meal and yeast. Levels in feed: Carp – 500-600 mg/Kg; Shrimp–400-2000 mg/Kg; Salmon and trout–800 mg/Kg.

Folic Acid

Deficiency symptoms – Anorexia, poor growth and lethargy. *Sources* – dried distillers solubles, cotton seed meal, rice bran, soybean meal and yeast. Levels in feed: Carp – 15 mg/Kg; Shrimp–10 mg/Kg; Salmon and trout–5-10 mg/Kg.

Cyanocobalamine (B_{12})

Deficiency symptoms – Anorexia and poor growth. *Sources* – blood meal, crab meal, fish solubles, fish meal. Levels in feed: Carp – 6 mg/Kg; Shrimp–0.1 mg/Kg; Salmon and trout–0.01-0.002 mg/Kg.

Ascorbic Acid

Deficiency symptoms – Black death (a disease characterized by melanized haemocytic lesion in collagenous layer) problems or decreased rate of molt, light coloured hepatopancreas, reduced growth and mortality. The Ascorbic acid in stored feeds is approximately 50 per cent/month. Coated forms of ascorbic acid (silicone, gelatin) are marginally available. Destruction of coated ascorbic acid during feed pelleting and extrusion would range from 30-50 per cent and 50-80 per cent respectively. *Sources* – Citrus fruits, liver, kidney tissues, fish tissues, goose berry (amla). Levels in feed: Carp – 30-50 mg/Kg; Shrimp–50-80 mg/Kg; Salmon and trout–300 mg/Kg.

Fat Soluble Vitamins

Vitamin D

Vitamin D_2 (ergocalciferal) and Vitamin D_3 (Cholecalciferal). Deficiency symptoms – poor growth, soft exoskeleton and lethargy. *Sources* – Fish liver oils, liver meals and fish meal. Levels in feed: Carp – 2000 IU/Kg; Shrimp–5000 IU/Kg; Salmon and trout–1000-2000 IU/Kg.

Vitamin A

It occurs in two forms – Vitamin A_1 (retinol) found in marine mammal and marine fish and Vitamin A_2 (retinol 2) found in freshwater fish. Deficiency symptoms – depigmentation and soft exoskeleton. *Sources* – Fish liver oils and liver meals. Levels in feed: Shrimp–10,000 IU/Kg.

Vitamin E

Fat soluble, antioxidant, tocopherols. *Sources*: dried distillers solubles, cotton seed meal, rice bran and wheat products. Levels in feed: Carp – 100 mg/Kg; Shrimp–300 mg/Kg; Salmon and trout–30-50 mg/Kg.

Vitamin K

Naturally occurs in two forms – K_1 in plants and K_2 in microorganism. Synthetic forms – K_3 (menadione) required for normal blood coagulation in animals. *Sources*: Liver meal and fish meal. Levels in feed: Carp – 4 mg/Kg; Shrimp–5 mg/Kg; Salmon and trout–10 mg/Kg.

The levels of vitamins in feeds are mostly reported in terms of milligrams of vitamin per kilogram of feed (mg/kg). Levels of vitamins A, D and E are, however, reported in terms of international units of activity (I.U.). These units are defined in the following table as follows:

International Units of Activity of Vitamins

Vitamin	*One I.U. (International Unit) Equals*
A	The activity of 0.344 microns of all-trans-vitamin A_1 acetate or 0.3 microns of retinal (vitamin A) or 0.6 microns of b -carotene
D	The activity of 0.025 microns of vitamin D_3 (cholecalciferol)
E	The activity of 1 mg of synthetic vitamin E acetate (dl-a -tocopherol acetate)

Source: Kutsky, 1981.

Vitamin E has anti-oxidative properties and may be required at higher levels in fish and shrimp diets which are high in PUFAs as these are susceptible to rancidity.

Characteristics of the Major Vitamins Important to Fish and Shrimp

Vitamin Synonyms	*Group 1*	*Solubility 2*	*Major Natural Sources*
Thiamine; Aneurine; B_1	B	W	legumes, brans, yeast
Riboflavin B_2	B	W	yeast, liver, milk, soybeans
Pyridoxine B_6	B	W	yeast, cereals, liver
Pantothenic acid	B	W	brans, yeast, animal offal; fish flesh
Niacin; nicotinic acid; niacinamide	B	W	yeast, legumes, forage
Biotin	B	W	liver; yeast; milk products

Contd...

Contd...

Vitamin Synonyms	*Group 1*	*Solubility 2*	*Major Natural Sources*
Folic acid; folacin	B	W	yeast; fish tissue and viscera; leaf meal, fish meal and viscera
Cyanocobalamin; APF 3; B_{12}	B	W	slaughterhouse wastes
Choline		W	wheat germ; legumes.
Inositol		W	legumes; yeast; wheat germ
Ascorbic acid* C		W	citrus fruits; fresh fish tissue; insects
Retinol A 4/		F	fish oils 5/
Cholecalciferol D 4/		F	fish oils
Tocopherols* E 2/		F	vegetable oils
Menadione K 4/		F	leaf meals

1/B = members of the B group of vitamins

2/W = water soluble vitamins. F = fat soluble vitamins

3/Animal protein factor

4/Hypervitaminosis (problems associated with excess dietary inclusion) can occur with these vitamins

5/Some carotenoids (*e.g.* b-carotene; astaxanthin) are precursors of vitamin A utilizable by some species

* These vitamins are particularly susceptible to losses in potency caused by heat, rancidity or feed processing. The water soluble vitamins also tend to leach into the aquatic environment.

Minerals (Micronutrients)

Minerals are a diverse group of inorganic compounds required in considerable and lesser quantities for essential functions in the body. Micro-minerals (trace minerals) are required in small amounts as components in enzyme and hormone systems. Fish can absorb many minerals directly from the water through their gills and skin, allowing them to compensate to some extent for mineral deficiencies in their diet. Mineral elements are important in many aspects of fish and shrimp metabolism.

Functions of Minerals

- ☆ Constituents of the exoskeleton (provide strength and rigidity to bones in fish and the exoskeleton of crustaceans).

- ✰ Balance of osmotic pressure (involved in body fluids, mainly with the maintenance of osmotic equilibrium with the aquatic environment and in the nervous and endocrine systems).
- ✰ Structural constituents of tissues (components of enzymes, blood pigments and other organic compounds).
- ✰ Essentially involved in the metabolic processes concerned with energy transport.
- ✰ Transmission of nerve impulses.
- ✰ Muscle contractions.
- ✰ Serve as essential components for enzymes, vitamins, hormones, pigments.
- ✰ Serve as co-factors in metabolism, catalyst and enzyme activators.

Classification

Minerals can be divided into two groups (macro-minerals and micro-minerals) based on the quantity required in the diet and the amount present in fish. These minerals regulate osmotic balance and aid in bone formation and integrity.

Minerals (Micronutrients)

Macro minerals (major minerals) – Required in considerable quantities. Most of the seven major 'minerals'–Calcium (Ca), Phosphorus (P), Potassium (K), Sodium (Na), Chlorine (Cl), Magnesium (Mg) and Sulphur (S) reported essential for terrestrial animal life are also believed to be required by fish. However, only seven, (Ca, P, Mg, Fe, Zn, I, and Se) have been shown to be required or utilized by salmonids.

Micro minerals (minor or trace minerals)- Required in lesser amounts or trace amounts fifteen trace elements–Iron (Fe), Zinc (Zn), Copper (Cu), Manganese (Mn), Nickel (Ni), Cobalt (Co), Molybdenum (Mo), Selenium (Se). Chromium (Cr), Iodine (I), Fluorine (F), Tin (Sn), Silicon (Si), Vanadium (Va), and Arsenic (As) reported essential for terrestrial animal life are also believed to be required by fish.

It can be assumed that the following elements at least are also essential for body functions: Na, Mo, Cl, Mn, Co, and probably Cr and F.

Fish and crustaceans can absorb minerals by other routes than from the digestion of food through the ingestion of seawater and through exchange from their aquatic environment across body tissues such as skin and the gill membranes. Therefore, the dietary requirements of minerals are largely dependent on the mineral concentration of the environment in which the fish and prawns are cultivated.

It is generally believed that Ca, Na, K and Cl requirement might be satisfied through absorption from the environment. Phosphorus is deficient in water and thus essential in the diet. Minerals are therefore probably not so important a component of the diet of fish and shrimp to that of other animals.

Calcium

	Requirement (g/kg of feed)
Indian major carps	5-18
Grouper	0.5
Trout and salmon	0.2-0.3
Common carp	0.2
Red sea bream	3.4

Calcium in the feed needs to be monitored to maintain a Ca:P ratio of 1:1 to 16:1.

Phosphorus

	Requirement (g/kg of feed)
Indian major carps	5-7
Channel cat fish	4-7
Trout and salmon	7-8
Common carp	6-7
Red sea bream, sea bass	7-8
Prawn	9
Shrimp	0.8 (available P); 1.5 (total P)
Tilapia	9

Sources: Dried distillers solubles, cottonseed meal, crab meal, fish solubles, fish meal, krill meal, rice bran, shrimp meal, squid meal, wheat bran and yeast.

Calcium is absorbed by fish from seawater; but, freshwater is low in calcium. However, since most feeding stuffs, particularly animal proteins, have high levels of calcium, its deficiency in fish through dietary insufficiency is most unlikely. On the other hand, both seawater and freshwater contain very little phosphorus and so this element is important from the dietary point of view. The level of phosphorus in feeds and feed components is therefore very important. Some types of phosphorus are unavailable to fish and an assessment of the availability of phosphorus in the diet is essential. Generally, animal sources of phosphorus are best absorbed by fish but some species, such as carp, do not absorb the element well from this source. Inorganic sources of phosphorus vary in their availability. But, some in common use in feedstuffs are mostly high in availability. The phosphorus from plant sources is generally poorly available. It should be noted that most feed compositional tables show levels of total phosphorus. Those that show a level for available phosphorus are designed for the manufacturers of poultry feeds and should not be applied to fish. The availability of the phosphorus in animal protein is taken as 100 per cent for poultry for example. There appear to be differences in the availability of phosphorus in the diet to various species of fish. So it is difficult to generalize. However, guidelines are suggested in following table:

Provisional Availability Factors for Phosphorus

Type of Ingredient	*Per cent Availability Factor (to Apply to the Total Phosphorus Level in the Feed)*	
	Stomachless Fish (eg., Cyprinids–Carp, etc.)	*Other Fish and Shrimp*
Plants and plant products	30	30
Animal products	30	70
Microbial products	90	90
Inorganic phosphorus		
Monobasic sodium, potassium	95	95
Calcium phosphates	95	95
Dibasic calcium phosphate	45	70
Tribasic calcium phosphate	15	65

There is some evidence that, although the actual level of calcium may not be important, the Ca/P ratio may be of importance in the nutrition of some fish and crustaceans.

Magnesium

	Requirement (g/kg of feed)
Carp	0.4-0.5
Trout and salmon	0.5-0.7
Prawn	0.8-1
Shrimp (per cent)	0.2

Sources: Cotton seed meal, crab meal, krill meal, rice bran, shrimp meal, wheat bran.

Sodium, Potassium and Chlorine

	Na	*K*	*Cl*
Fish	1-3	1-3	1-5
Prawn	6	9	–
Shrimp (per cent)	0.6	0.9	–

Sources: Crab meal, fish solubles, fish meal, krill meal, shrimp meal.

Sulphur

	Requirement (g/kg of feed)
Fish	3-5
Prawn	0.2

Sources: Fish meal, cotton seed meal, rape seed meal and yeast.

Iron

	Requirement (g/kg of feed)
Fish (carp)	150
Prawn	5-20
Shrimp (ppm)	300

Sources: Blood meal, crab meal, fish meal, fish solubles.

Copper

	Requirement (g/kg of feed)
Trout and salmon	3
Channel cat fish	1.5
Fish	1-4
Prawn	25
Shrimp (ppm)	35

Sources: Dried distillers solubles, fish solubles, krill meal and yeast.

Zinc

	Requirement (g/kg of feed)
Trout and salmon	15-30
Fish	30-100
Prawn	50-100
Shrimp (ppm)	110

Sources: Dried distillers solubles, corn gluten meal, fish solubles, krill meal, rice bran, wheat bran and yeast.

Manganese

	Requirement (g/kg of feed)
Trout and salmon	12-13
Carp	4
Shrimp (ppm)	20

Sources: Dried distillers solubles, corn gluten meal, fish solubles, krill meal, rice bran, wheat bran and yeast.

Selenium

	Requirement (g/kg of feed)
Trout and salmon	0.1-0.4
Prawn	1
Shrimp (ppm)	1

Sources: Blood meal, corn gluten meal, fish solubles, fish meal and yeast.

Cobalt

	Requirement (g/kg of feed)
Prawn	10
Shrimp (ppm)	10
Fish	5-10

Sources: Cotton seed meal, soybean meal, fish meal and yeast.

A summary of the available information on the mineral requirements of fish is given in following table.

Minerals	*Dietary Requirements*
Ca	0.5 per cent
P (available)	0.7 per cent
Mg	0.05 per cent
Na	0.1-0.3 per cent
K	0.1-0.3 per cent
S	0.3-0.5 per cent
Cl	0.1-0.5 per cent
Fe	50-100 mg/kg
Cu	1-4 mg/kg
Mn	20-50 mg/kg
Co	5-10 mg/kg
Zn	30-100 mg/kg
I	100-300 mg/kg
Mo	trace
Cr	trace
F	trace

Source: ADCP, 1983.

Examples of Vitamins and Mineral Mixes

The following appendix contains the composition of the premixes used in the formulae are:

Mineral Mix No. 1 (Trout, Carp, Tilapia and Catfish)

Ingredient	*mg/g of Premix*[1/]
Fe	50
Cu	3
Co	0.01
Mn	20
Zn	30
I	0.1
Se	0.1

Source: Chow, 1982c.

1/: To be used at 0.1 per cent in diet.

Mineral Mix No. 2 (Catfish) 2/2/*Siluris* sp.

Ingredient	*Amount (g)*[3/]
Calcium Carbonate	150.0
$MnSO_4$ H_2O	15.0
$ZnSO_4$ $7H_2O$	35.0
$CuSO_4$ $5H_2O$	2.0
$FeSO_4$ $7H_2O$	25.0
KI or KIO_4	0.1
Na H_2 PO_4	1000.0
$MgSO_4$	50.0
NaCl	223.0
	1500.1

Source: Halver, 1982.

3/: Premix to be used at 1.2 per cent in moist diet.

Mineral Mix No. 3 (Trout)

Ingredient	*g/kg of Premix*[1/]
$ZnSO_4$	185.10
$FeSO_4$ $7H_2O$	49.60
$CuSO_4$	3.86
$MnSO_4$	207.20
KIO_3	0.84
Inert Carrier	553.40
	1 000.00

Source: Piper *et al.*, 1982

1/: To be used at 0.1 per cent in diet.

Mineral Mix No. 4 (Indian Carps)

Ingredient	*ppm in Final Diet*[2/]
$CuSO_4$ $5H_2O$	Cu = 10
$FeSO_4$ $7H_2O$	Fe = 100
$MnSO_4$ H_2O	Mn = 50
ZnO	Zn = 50
$CaCl_2$ $6H_2O$	Co = 0.05
KI	I = 0.1
$CaHPO_4$	(Filler)

Source: Chow, 1982.

2/: When mineral mix used at 0.1 per cent of diet.

Mineral Mix No. 5 (Tilapia)

Ingredient	*For Use for Diets for Fish in Freshwater (g/kg premix)*[3/]	*For Use for Diets for Fish in Seawater (g/kg premix)*[4/]
$CaHPO_4$ $2H_2O$	727.7775	
$MgSO_4$ $7H_2O$	127.5000	510.00
NaCl	60.0000	200.00
KCl	50.0000	151.11
$FeSO_4$ $7H_2O$	25.0000	100.00
$ZnSO_4$ $7H_2O$	5.5000	22.00
$MnSO_4$ H_2O	2.5375	10.15
$CuSO_4$ $7H_2O$	0.7850	3.14
$CoSO_4$ $7H_2O$	0.4775	1.91
$Ca(IO_3)$ $6H_2O$	0.2950	1.18
$CrCl_3$ $6H_2O$	0.1275	0.51
	1000.0000 g	1000.00 g

Source: Jauncey and Ross, 1982.

3/: To be used at 4 per cent in diet

4/: To be used at 1 per cent in diet.

Vitamin (and Mineral) Mix No. 1 (Warm-Water Fish)

Ingredient	*mg/kg of Dry Diet*
Vitamin A	6 000 I.U.
Vitamin D_3	1 000 I.U.
Vitamin E	60 I.U.
Vitamin K	12
Vitamin C	240
Vitamin B_1	24
Vitamin B_2	24
Panthothenic acid	60
Niacin	120
Vitamin B_6	24
Biotin	0.24
Folic acid	6
Choline Chloride	540
Vitamin B_{12}	0.024
Fe	50
Cu	3
Mn	20
Zn	30
I	0.1
Co	0.01
Se	0.1

Source: Chow, 1982a.

Vitamin (and Mineral) Mix No. 2
(Rabbit Fish, Sea Bass and Grouper)

Ingredient	*mg/kg of Dry Diet*
Thiamin HCl	40
Vitamin B_2	40
Pyridoxine HCl	40
Nicotinic Acid	150

Contd...

Contd...

Ingredient	*mg/kg of Dry Diet*
Calcium Pantothenate	100
Folic Acid	5
Biotin	1
Vitamin B_{12}	0.02
Inositol	800
Choline Chloride	3 500
Sodium Ascorbate	2 000
Vitamin E	200
Vitamin K	80
Vitamin A	5 000 I.U.
Vitamin D	1 000 I.U.
Zn	40
Mn	20
Cu	4
I	0.8
Co	0.12

Vitamin Mix No. 3 (Freshwater Prawns, Marine Shrimp and Sea Bass)

Ingredient	*mg/g of Premix*[1/]
Vitamin A	500 I.U.
Vitamin D_3	100 I.U.
Vitamin B_1	0.1
Vitamin B_2	0.3
Pyridoxine	0.2
Vitamin B_{12}	0.001
Nicotinic Acid	2.0
Calcium Pantothenate	0.6
Folic Acid	0.05
Vitamin K	0.2
Vitamin C	5.0

Source: Chow, 1984.

1/: Used at various amounts in diet according to species.

Vitamin Mix No. 4 (Marine Shrimp)

Ingredient	*mg/kg of Dry Diet*
Thiamin HCl	120
Vitamin B_2	40
Pyridoxine HCl	120
Nicotinic Acid	150
Calcium Pantothenate	100
Folic Acid	5
Biotin	1
Vitamin B_{12}	0.02
Inositol	4 000
Choline Chloride	1 200
Sodium Ascorbate (Vit. C)	5 000
Vitamin E	200
Vitamin K	40
Vitamin A	5 000 I.U.
Vitamin D	1 000 I.U.

Source: Kanazawa, 1984.

Vitamin Mix No. 5 (Trout, Carp, Tilapia and Catfish)

Ingredient	*mg/g of Premix*[1/]
Vitamin A	1 000 I.U.
Vitamin D_3	200 I.U.
Vitamin E	10 I.U.
Vitamin K	2
Vitamin B_1	4
Vitamin B_2	4
Pantothenic Acid	10
Niacin	20
Pyridoxine	4
Biotin	0.02
Folic Acid	1
Vitamin C	40
Choline Chloride	90
Vitamin B	0.004
Ethoxyquin	16

Source: Chow, 1982 c.

1/: To be used at 0.6 per cent in the diet.

Vitamin Mix No. 6 (Catfish) 2/2/*Silurus* sp.

Ingredient	*mg/kg Premix*[3/]
Vitamin A	100 000 I.U.
Vitamin D_3	50 000 I.U.
Vitamin E	1 000 I.U.
Vitamin K (Menadione)	500
Vitamin C	10 000
Biotin	50
Choline	150 000
Folic Acid	250
Niacin	77 500
Pantotheic Acid	2 000
Pyridoxine	500
Vitamin B_2	1 000
Vitamin B_1	500
Inositol	1 500
Vitamin B_{12}	1
BHT or Ethoxyquin	7 500 or 5 000
Add Maize or Wheat	to 1 000 g

Source: Halver, 1982.

3/: To be used at 0.8 per cent in the diet.

Vitamin Mix Nos. 7 (Oregon Salmon) 1/and 8 (Trout)

Ingredient	*Amount per kg of Premix*	
	No.7[2/]	*No.8*[3/]
Vitamin A	-	1 653 000I.U.
Vitamin D	-	110 200I.U.
Vitamin E	33 501 I.U.	88 160 I.U.
Vitamin K (sodium bisulphite)	1 201.12 mg	2 755.00 mg
Vitamin C	59.51 g	165.30 g

Contd...

Contd...

Ingredient	*Amount per kg of Premix*	
	No.7[2/]	*No.8*[3/]
Biotin	39.67 mg	88.20 mg
Vitamin B	3.97 mg	5.51 mg
Folic Acid	848.50 mg	2 204.00 mg
Inositol	17.63 g	g
Niacin	12.56 g	55.10 g
D-Calcium pantothenate	7.05 g	26.45 g
Pyridoxine HCl	1 179.10 mg	7 714.00 mg
Vitamin B_2	3.53 g	13.22 g
Vitamin B_1 (mono nitrate)	1 714.70 mg	8 816.00 mg
Add Cereal Carrier	[to 1 000.0 g]	[to 1 000.0 g]

Source: Piper *et al.,* 1982.

1/: Pacific salmon.

2/: To be used at 1.5 per cent in diet.

3/: To be used at 0.4 per cent in diet.

Vitamin Mix No. 9 (Indian Carps)

Ingredient	*mg/kg of Diet*[4/]
Vitamin A	5 000 I.U.
Vitamin D	600 I.U.
Vitamin B_1	10
Vitamin B_2	20
Pantothenic Acid	30
Niacin	50
Vitamin C	200

Source: Chow, 1982a

4/: When vitamin mix used at 0.1 per cent of diet.

Vitamin Mix No. 10 (Tilapia)

Ingredient	*Amount in g/kg Premix*[1/]
Vitamin B_1	2.5
Vitamin B_2	2.5
Vitamin B_6	2.0
Pantothenic Acid	5.0
Inositol	100.0
Biotin	0.3
Folic Acid	0.75
Para aminobenzoic Acid	2.5
Choline	200.0
Niacin	10.0
Vitamin B_{12}	0.005
Vitamin A	100 000 I.U.
Vitamin E	20.1
Vitamin K	2.0
Vitamin C	50.0
Vitamin D_3	500 000 I.U.

Source: Jauncey and Ross, 1982.

1/: To be used at 2 per cent in the diet.

Vitamin Mix No. 11 (Marine Percidae) 2/(Provisional) 2/Yellowtail, sea bass, sea bream, grouper

Ingredient	*mg/kg Dry Diet*
Vitamin A	6 000 I.U.
Vitamin B_1	20
Vitamin B_2	20
Vitamin B_6	20
Vitamin B_{12}	0.02
Folic Acid	5
Inositol	600
Niacin	150
Pantothenic Acid	50

Contd...

Contd...

Ingredient	*mg/kg Dry Diet*
Vitamin C	200
Choline	2 000
Vitamin D	2 500 I.U.
Vitamin E	200
Biotin	1
Vitamin K	10

Source: New, 1986a.

Vitamin Mix No. 12 (Sea bass, grouper and siganids)

Ingredient	*mg/kg Dry Diet*
Vitamin A	5 000 I.U.
Vitamin B_1	120
Vitamin B_2 (Thiamin HCl)	40
Vitamin B_6 (Pyridoxine HCl) H_2O	120
Vitamin B_{12}	0.02
Folic acid	5
Inositol	800
Niacin	150
Calcium pantothenate	100
Vitamin C (Sodium ascorbate)	1 000
Choline HCl	1 200
Vitamin D	1 000 I.U.
Vitamin E (Tocopherol)	200
Biotin	1
Vitamin K. (Menadione)	40

Source: Meyers, 1987b.

Vitamin/Trace Element Mix No. 13 (Common Carp)

Ingredient	*mg/kg Pelleted Feed*
Vitamin A	8 000 I.U.
Vitamin D	900 I.U.
Vitamin E	2 I.U.
Vitamin K	4
Vitamin B_2	3.6
Niacin	20
Choline Chloride	160
Pantothenic Acid	7
Vitamin B_6	0.2
Vitamin B_{12}	0.005
Mn	70
Zn	60
Fe	20
Cu	2
I	1
Co	0.2

Source: Viola *et al.*, 1982.

Vitamin (and Mineral) Mix No. 14 (Freshwater Prawns)

Vitamin Mix	*Amount/kg diet*	*Mineral Mix Amount/kg diet (mg)*	
Ingredient		*Ingredient*	
Vitamin A	5 500.0 I.U.	Zinc oxide	55.1
Vitamin D	1 237.0 I.U.	Ferrous sulphate and carbonate	59.5
Vitamin E	4.1 I.U.		
Vitamin K	0.8 mg	Manganous oxide	56.0
Vitamin B_2	3.3 mg	Ethylenediaminedihydroiodide	0.25
Pantothenic acid	4.9 mg		
Niacin	24.7 mg	Cobalt sulphate	0.50
Choline chloride	67.1 mg	Sodium chloride	2 646.0
Vitamin B_{12}	8.2 mg	Copper oxide	4.5
Folic acid	0.3 mg	Sodium selenite	0.10

Source: Corbin *et al.*, 1983.

4/: The total mineral content of feedstuffs is also referred to as the 'ash content' because of the method of analysis.

Non-Nutrient Dietary Components

It is otherwise called as non-toxic dietary components. These components do not have any apparent nutritive value. Two major components are concerned and they are Fibre and Ash.

Fibre

It refers to mixtures of cellulose, hemi cellulose, lignin, pentasans and other undigestible fractions in the feed. Finfish do not have capacity to digest fibrous material such as cellulose and lignin. Certain amount of fibre in feeds permits better binding and moderates the passage of food through the alimentary canal. However, it is not desirable to have a fibre content exceeding 8-12 per cent in diets for fish; 4 per cent for shrimp. If the fiber content is excessive, it results in a decrease in the total dry matter and nutrient digestibility of the diet resulting in poor performance. Suitable plant material can be included in diets as fillers and this is often done in experimental studies. Chitin in shrimp feed is believed to have a growth promoting effect (0.5 per cent in shrimp feed). Shrimp meal is a good source of chitin.

Ash

Ash is a heterogeneous group of materials including the non-combustible inorganic components of feed stuffs and diets. Ash may also include contaminants such as silica (sand grains). It is not desirable for the ash content to exceed 12 per cent.

Like fibre, the ash contents affect the digestibility of diets. Ash refers to inorganic elements (minerals) in the feed.

Chapter 9

Nutritional Bioenergetics of Fish

Energy

Fish and shrimp require food to supply the energy that they need for movement and all the other activities in which they engage and the 'building blocks' for growth. In this, they do not differ from other farm animals, or humans. However, aquatic animals are 'cold-blooded'. Their body temperature is the same as the water in which they are living. They do not, therefore, have to consume energy to maintain a steady body temperature and they tend to be more efficient users of food than other farm animals. Their metabolic rate, however, depends very largely on the temperature of the water in which they are living. The optimum temperature (that at which they will grow best) is different for each species. Within the range of temperatures of which they are tolerant (those at which they will survive, eat and grow) metabolic rate. and the need for food, increases as the optimum temperature is reached. Thus, in areas where there is a wide range of water temperature seasonally, fish will eat much more food in the summer than in the winter.

Energy can be defined as the capacity to do work. Energy is required to do mechanical work (for example, muscle activity for movement), chemical work (the chemical processes which take place in the body), electrical work (nerve activity) and osmotic work (maintaining the body fluids in an equilibrium with each other and with the medium, whether fresh, brackish or seawater in which the animal lives). Free energy is that which is left available for biological activity and growth after the energy requirements for maintaining body temperature (not necessary for fish) is satisfied. Excess energy is dissipated as heat.

From the point of view of the fish or shrimp farmer, the most economically important thing is the quantity and cost of the energy which is available for the growth of the animal being cultured. Food supplies this energy. The food requirements of different species vary in quantity and quality according to the nature of the animal, feeding habits, size, environment, reproductive state, etc.

The gross energy (or gross calorific value) of a food, sometimes designated as GE, is the total energy contained in it. Different components of the diet have different energy availabilities. The digestible energy (DE) of a food is the GE of the food less the energy of the faeces excreted.

Metabolism is the sum of all the chemical and energy transactions of the body. It is the process by which nutritive material is built into living matter. Metabolism includes the storage of energy (anabolism) as fat, protein and carbohydrate, and its conversion (catabolism) into free energy for work and growth.

The metabolic rate of small fish and shrimp is greater than that of large animals. Small animals grow faster than large ones in terms of percentage of increase in weight per day. Thus, the feed requirements of small fish and shrimp are different to those of larger animals; small animals require a higher feeding rate. At a certain body size, growth rate starts to decline rapidly. The optimum marketable size of an aquaculture species normally occurs at this point, unless market factors dictate otherwise.

Fish Nutritional Bioenergetics

Study of balance between energy intake (in the form food) and energy utilization (by animals) for life sustaining processes such as maintenance, activity and tissue synthesis is called Fish Nutritional Bioenergetics. Rate of transfer can be expressed as dB/dt. Also known as energy budget.

C= F+U+M+delta B – simplified equation,

where,

C: Amount of food energy consumed (also known as ration)

F: That part lost as faeces (FE)

U: Non faecal nitrogenous loss (UE + GE or ZE)

M: Loss by way of metabolism (ME)

B: Change in materials of body growth (RE+HE+SE)

In other words C *i.e.* IE = FE+ME + (UE+GE) +HE+RE+SE).

Basic unit of heat = calories *i.e.* gcal. Calories–amount of heat required to raise the temperature of 1 g of water to 1°C, measured from 14.5-15.5° C; 1 joule =0.239 calories or 1 gcal = 4.184 joules.

Types of Energy

- **Gross energy (E)** – energy that is released as heat when a substance is completely oxidized to CO_2, NO_2 and other gases (heat of combustion).
- **Intake energy (IE)**– gross energy consumed by animal in its food. Majority of IE in the form of CHO, Protein and fat
- **Faecal energy (FE)**–gross energy of the faeces.
- **Digestible energy (DE)** = Intake energy – faecal energy (IE-FE)
- **Urinary energy (UE)**–gross energy in urinary products.
- **Gill excretion energy (GE or ZE)**–gross energy of the compounds excreted through the gills of aquatic animals
- **Surface energy (SE)** – energy lost from the surface of an organism
- **Metabolizable energy (ME)** = *i.e.* ME = IE- (FE+UE+ZE or GE) or ME = DE – (GE+UE)
- **Retained energy** (RE = ME – HE)–that portion of energy contained in the feed that is retained as part of the body or voided as a useful product such as gametes.

Energy flow pattern or typical path ways of utilization of food energy.

Intake energy (C = 100 per cent)

→ Faeces (F = 25 per cent)

Digestible energy (75 per cent) *i.e.* IE-FE (100-25)

→ Urine and gill excretion (8 per cent)

Metabolizable energy ME = IE – (FE+UE+GE) *i.e.* 100 – (25+8)
(67 per cent)
(or) DE – (GE + UE) *i.e.* 75-(8) = 67

Heat energy (33 per cent)	Retained energy (ME-HE) *i.e.* 34 per cent
Basal metabolism	Lactation energy
Voluntary activity	Tissue energy
Product formation	Ovum energy
Digestibility and absorption	Conceptus energy
Fermentation	Epidermal energy
Thermal regulation	
Water formation and excretion	

Methodology of Fish Nutritional Bioenergetics

A) Food consumption

Done by two methods:

1. **Direct method** (Food offered-food left = food consumed)
2. **Indirect method** – by consumption *i.e.* F+U+M+ delta B; markers method (Use of non assimilable markers and use of radioisotopes)

Use of Radioisotopes

Food is labeled with one of the following isotopes. ^{14}C, ^{32}P, ^{45}Ca, ^{35}S. If it is used for phytoplankton feeders, feed algae is grown in medium containing isotopes. If used for formulated diets, the isotope is mixed with feed. ^{32}P is very convenient isotope having short half life of 14.2 days and radiation could be easily detected and measured in small animals. But, due to its short half life, it cannot be used for experiments of longer duration. ^{14}C has the advantage of a half life of 5760 yrs.

Use of Non-assimilable Markers

Most commonly used marker is Cr_2O_3 which is well mixed at 5 per cent level with the feed and its exact quantity in the feed and faeces is chemically estimated.

B) Assimilation

Two methods (Direct and Indirect).

Direct Method

(Assimilation = Consumption–Faeces)

Indirect Methods

$$A = \text{Consumption} - \frac{\text{Conc. of indicator in food/unit weight}}{\text{Conc. of indicator in faeces/unit weight}}$$

Assimilation Efficiency = Assimilation% Consumption × 100

C) Metabolism

Two methods (direct and indirect methods).

Direct Method

By using static or flow through respirometers, the metabolism (in terms of O_2 consumption) of fish fed with the test diet is estimated.

Indirect Methods

(M = C – (F+U+ delta B)

where,

C: Amount of food energy consumed (also known as ration)

F: That part lost as faeces (FE)

U: Non faecal nitrogenous loss (UE + GE or ZE)

M: Loss by way of metabolism (ME)

B: Change in materials of body growth (RE+HE+SE)

D) Specific Dynamic Action (SDA)

It denotes the energy cost of biochemical transformation of ingested food in to a metabolizable excretable form (Heat increment during assimilation). SDA is lower for fishes when compared to higher vertebrates.

$$SDA = M - (Ms + Me)$$

where,

M: Metabolic rate of fed animals

Ms: Metabolic rate of starved animals

Me: The exited metabolic rate due to feeding procedure.

E) Growth

At the end of experiments, the animals are weighed, measured and dried in an oven.

It is better to determine the proximate composition which would be of use in interpretation of the result.

Analysis – Formula

Gross Conversion Efficiency (K1)

$$\frac{\text{Production}}{\text{Consumption}} \times 100$$

Net Conversion Efficiency (K2)

$$\frac{\text{Production}}{\text{Assimilation}} \times 100$$

Trophic Coefficient

$$\frac{\text{Consumption of feed (in g dry wt)}}{\text{Production of flesh (in g dry wt)}}$$

Average Daily Growth

$$\frac{\text{Present weight} - \text{Previous weight}}{\text{Time interval of both weighing days}}$$

FCR (Food Conversion Ratio)

$$\frac{\text{Food consumed}}{\text{Weight gained}} \text{ or } \frac{\text{Consumption}}{\left(\overline{\text{Wn}} + \text{D}\right) - \overline{\text{W0}}}$$

where,

$\overline{\text{W0}}$ = average initial weight

$\overline{\text{Wn}}$ = average final weight

D = weight of dead animals

C = total feed consumed

Apparent FCR

$$\frac{\text{Consumption (supplementary feed)}}{\text{Weight gain}}$$

True FCR

$$\frac{\text{Consumption (supplementary feed + natural feed)}}{\text{Weight gain}}$$

FCE (Food Conversion Efficiency)

It is reciprocal of FCR and it is expressed in per cent

$$\text{Food Conversion Efficiency} = \frac{\text{Weight gain}}{\text{Consumption}} \times 100$$

Specific Growth Rate (SGR)

$$\frac{\ln W_2 - \ln W_1}{t} \times 100$$

where,

ln: Natural log

W_2: Final weight

W_1: Initial Weight.

t: Time *i.e.* days

Estimation of Energy Value of Feed Ingredient – Two Methods

Chemical Composition

Gross energy fuel value for protein (5.6), fat (8-9.5) and CHO (4.1) is taken; Now, the feed is analysed for protein, fat, ash and fiber;

CHO is estimated as:

Dry Matter – (% of ash on dry basis + % of crude fibre on dry basis + % of crude fat on dry basis + % of crude protein on dry basis)

(or)

100 – (% of ash on dry basis + % of crude fibre on dry basis + % of crude fat on dry basis + % of crude protein on dry basis + per cent of moisture on wet basis)

The fuel value is calculated as follows:

$$\text{Fuel value} = (\text{Protein} \times 5.6) + (\text{Fat} \times 8\text{–}9.5) + (\text{CHO} \times 4.1)$$

Bomb Calorimetric Method

Principle

A Bomb Calorimeter is an apparatus used for measuring heats of combustion. The amount of heat measured in calories that is released when a substance is completely oxidized in a bomb calorimeter containing 25-30 atmosphere of O_2 is called gross energy of a substance.

Apparatus Required

Bomb calorimeter with firing unit, Pellet press, Oxygen charging device.

Procedure

Sample of feed to be tested is preweighed into combustion capsule

↓

Combustion capsule is placed in an oxygen bomb containing 25-30 atmosphere of O_2

↓

O_2 bomb is covered with 2000g of water in a calorimeter

↓

The bomb and calorimeter are adjacent to the same temperature

↓

Now, the sample is ignited with a fuse wire by cotton thread and temperature fire is measured

↓

Rinse inner surfaces of bomb with distilled water and collect all washings and titrate against $Na_2 CO_3$ using methyl orange indicator

↓

From the hydrothermal equivalent of calorimeter, the calories content of sample is calculated

Calculation

Gross energy (calorie/g)

$$= \frac{(FT - IT)\{(W) - (CV_T + CV_W) - (\text{ml of } Na_2\ CO_3 \ \text{consumed})\}}{\text{Weight of sample}}$$

where,

FT: Final temperature in firing unit

IT: Initial temperature in firing unit

CV_T: Calorific value of thread = 2.1/cm (when using thread 10 cm, CV of thread

= 2.1 × 10 = 21cal)

CV_W: Calorific value of ignition wire = 2.33/cm (when using wire 6 cm,

CV of wire: 2.33 × 4 = 9.32 cal)

W: Water equivalent or hydrothermal equivalent thermal calorimeter in calories per degrees centigrade.

Determing Water Equivalent (W)

The water equivalent of the system is determined by igniting a pellet of supplied pure and dry benzoic acid of known calorific value weighing not less than 0.8 and not more than 1.1 gram. Record the corrected temperature rise (T), calculate the calorific values of thread and wire (CV and CW) using the data given below and evaluate water equivalent by substitution in the equation. W is calculated by the following formula:

$$W = \frac{H \times M + (CV_T + CV_W)}{T}$$

where,

W: Water equivalent or hydrothermal equivalent thermal calorimeter in calories per degrees centigrade

H: Known calorific value of benzoic acid in cal/gram

T: Final temperature in firing unit–Initial temperature in firing unit

CV_T: Calorific value of thread = 2.1/cm (when using thread 10 cm, CV of thread

= 2.1 × 10 = 21cal)

CV_W: Calorific value of ignition wire = 2.33/cm (when using wire 6 cm,

CV of wire = 2.33 × 4 = 9.32 cal)

M: Mass of sample in grams

Notice: 4 cm wire is normally used because approximately 2 cm wire is used to strap around the electrodes. The thread burns completely, therefore the calorific value remains constant, which is 21 for 10 cm.

Observation

Standardization with a 0.955 gram benzoic acid sample (6319 cal/gram) produced a net corrected temperature rise of 2.609°C. Nichrome wire and cotton thread of 4 and 10 cm lengths were consumed in the firing.

$$W = \frac{H \times M + (CV + CV)}{T}$$

Substitution in the standardization equation:

H = 6319 cal/gram

M = 0.955 gram

CV = (10 cm) (2.1) = 21 cal.

CV = (4 cm) (2.33) = 9.32 cal.

T = 2.609

$$W = \frac{(6319)(0.955) + (21 + 9.32)}{2.609}$$

$$= 2325 \text{ cal/°C}$$

Factors Affecting Energy Partitioning

Basal metabolic rate–body size, oxygen availability, temperature, osmoregulation, stress, cycles.

Non-basal metabolic rate- gonadal growth and locomotion.

Factors Affecting the Energy Partitioning Basal Metabolic Rate

a) Body Size

Direct relation with metabolic rate. Metabolic rate increase directly with increasing body size. However, energy demand of a piece of tissue depends on size of the animal. This effect is called scaling and described mathematically by the formula:

$$Y = a\,X^b$$

where,

Y: Any physiological variable (Metabolic rate)

a: Proportionality constant

X: Body mass

b: Effect of size on Y

The value of 'b' is considered between 0.7 and 0.8. Hence, larger the fish, the less energy/unit body weight to maintain basal metabolic rate.

b) Oxygen Availability

Aquatic animals consume O_2 either at a rate directly dependent upon ambient oxygen tension (conformers) or at a rate independent of ambient oxygen tensions (Non-conformers or regulators). In conformers, metabolic rate is greater at higher O_2 tensions. In Non-conformers, O_2 consumption remains constant until O_2 tension in the ambient water reaches a critically low level. Therefore, advantage to operate the culture system with O_2 tensions in excess of critical point, ideally as near as possible to saturation.

Temperature

Direct relation with metabolic rate. Water temperature plays on extremely important role in energy partitioning. Two effects of temperature can be observed in aquatic animals. When an animal is acclimated to a certain temperature and they introduced to greater temperature, its metabolic rate will increase. If the animal is soon returned to the original temperature, its metabolic rate will return to the original rate. Temperature effect on metabolic rate can become quite significant.

Osmoregulation

Salinity of an environment of fish plays an important role in energetic cost of osmoregulation. Freshwater fish live in an environment *i.e.* hypo saline to their body tissues and hence exchange continually, water and sequester ions. Salt water fish have exactly opposite problems and expend energy availing the loss of water and in excreting ions.

Stress

Stress results in increased basal metabolic rate and induced by several factors. Accumulation of waste products in water, low $O_{2,}$ crowding, handling, external disturbances, water pollution, poor quality feed, hypoglycaemia phenomenon exhibited by fish cycles. All animals display cycling of their physiological processes. Some cycles are clear; but many cycles are much more subtle.

Factors Affecting Non-basal Metabolic Rate

Gonadal Growth

Gonadal growth results in the diversion of large amounts of

energy away from growth of muscle and other activities. At reproductive season, gonads may account for large proportion of body of a fish weighs, even up to 30 – 40 per cent.

Locomotion

Energy cost of locomotion is a major part of total energy consumption; varies between species depending on body shape and behavioural patterns. Energy expenditure on either gonadal development or locomotion is so great.

Self Assessment Questions

1. What is energy?
2. What is calorie?
3. What is heat?
4. What is catabolism, anabolism and metabolism?
5. What is bioenergetics?
6. What is fish nutritional bioenergetics?
7. What is DE, IE, FE?
8. List out the different types of energy.
9. What are the different methods of food consumption?
10. Write down the formula for FCR, FCE, SDA, SGR.
11. Collect different energy flow budgets for different fishes.
12. FCR of 1:3 means what?
13. FCE of 55 per cent means what?
14. What is scaling effect?
15. Mention the methods to determine energy for feed sample.
16. What are the factors affecting energy partitioning of fish?
17. Draw rough fish diagram and indicate all energy involved.
18. Difference between True and Apparent FCR.

Chapter 10

Aqua Feed Formulation and Management

Definition for Aqua Feed or Diet Formulation

Diet formulation is not an easy task. It is a process in which the appropriate feed ingredients are selected and blended to produce a diet with the required quantities of essential nutrients. Feed must be nutritionally balanced, pelletable, palatable and easy to store and use. The use of scientifically formulated and optimally processed aqua feed, definitely leads to sustainable production of fishes.

The importance aspects of aqua feed formulation and management are maximizing growth rate, increase production/unit area/unit time, increase reproductive efficiency and minimize mortality in aquaculture.

Food Constituents

For better survival, fish requires balance diet that is containing organic and inorganic ingredients that are referred as food constituents. According to their functions in the body, food constituents are classified as:

1. Body builders–Protein, minerals, salts
2. Energy producers – CHOs, fats.
3. Regulators – Vitamins.

Factors to be Considered in Diet Formulation

1. Nutrient requirements of species cultivated
2. A knowledge of nutrient and available energy
3. Digestibility and nutrient availability
4. Other dietary components
5. Dietary interactions: Micronutrient – diet composition interaction (Mineral – mineral interactions, Vitamin – mineral interactions, Vitamin – vitamin interactions.
6. Flavor quality

Factor 1: Nutrient Requirement of Species Cultivated

It is vital one and formulated feed meet the nutritional requirement of the cultured fish. Notable difference in nutritional requirement are seen in the essential fatty acids and energy. Essential amino acids requirement tend to remain constant both within and between species. Therefore, protein sources selected should meet these requirements.

Factor 2: Composition of Ingredients (in terms of nutrients and energy)

A knowledge of nutrient composition and available energy of dietary ingredients are essential for use in diet formulation. Most comprehensive information on feed composition is provided in U.S. Canada table of feed composition (National Research Council, NRC, 1983). The above values in NRC table are average one. The composition of feed ingredients is known to vary regionally, seasonally and also with the soil fertility and type of processing and storage. Therefore, it is desirable that each ingredient is analyzed for actual content prior to formulation. Further, feed ingredients also need to be screened for enzyme inhibitors and other indigenous toxins as well as aflatoxins and other mycotoxins.

Factor 3: Digestibility and Nutrient Availability

A knowledge of the digestibility of individual nutrients of all the ingredients is essential. The process of estimating their content is tedious and time consuming. However, general digestibility values for most of the ingredients for the widely cultured species are presently available and are often used in feed formulation.

	GE (Kj/g)	DE (Kj/g)
CHO		
Non legumes	17.15	12.55
Legumes	17.15	8.37
Protein		
Plant	23.01	15.90
Animal	23.01	17.78
Fat	38.07	33.47

Factor 4: Other Dietary Compounds

Other than main ingredients, feed additives are added.

Factor 5: Dietary Interactions

Four main types of nutrient interactions are known in finfish. They are:

1. Micro nutrient – dietary composition interaction.
2. Mineral – Mineral interactions
3. Vitamin – mineral interactions
4. Vitamin – Vitamin interactions

The above four interactions are influenced by a number of factors such as diet composition, diet processing, species of the cultured organism, age and environmental factors.

Nutrient Interaction 1: Micronutrient – Diet Composition Interactions

Thiamine – Sparing effect with high fat diets:

Thiamine (B_{12}) availability is known to be influenced by fat and protein content of diets of same calorific value. An experiment conducted on trout (*Oncorynchus mykiss)* at 15°C indicated that when fed Thiamine – deficient with high CHO diets, trout develop thiamin deficiency and showed higher mortality than those fed with thiamin deficient high fat diet. This is an indicative of a Thiamine – sparing effect with high fat diets. Similarly, pyridoxine (B_6) is also related to dietary protein metabolism.

Nutrient Interaction 2: Mineral – Mineral Interactions

A number of interactions between essential minerals are known to coexist. Magnesium requirement is dependent on Calcium and Phosphorus content of diet. Copper and Zinc may be antagonist. However, Magnesium requirement is not known to increase when dietary Calcium and Phosphorus is increased. As a result of this antagonism, these two minerals may compete for binding sites on protein.

Nutrient Interaction 3: Vitamin – Mineral Interactions

Several vitamin-mineral interactions in fish have been reported. There are two major interactions in fish. They are, Ascorbic acid–Iron and Ascorbic acid–Copper. Ascorbic acid is involved in metabolism of iron in fish. Therefore, Vitamin C deficiency results in reduction in serum iron and haemoglobin.

Nutrient Interaction 4: Vitamin – Vitamin Interaction

Vitamin B_{12} and Folic acid interaction in *L. rohita* has been reported well. When both vitamins are deficient, appearance of anemia was accelerated.

Flavor Quality

Presence off-flavor in mud due to chemical compound called geosmin. It is produced by fungus group *Actinomycetes* and blue green alga (*Ocillatoria* spp). Other source of off flavor is from industrial waste. The most chemicals involved are phenols, tars, and mineral oils. Depuration process must be done in freshwater fishes especially for carps to remove the earthy odour.

Methods Used for Feed Formulation

Introduction

To formulate an aqua feed, the proximate composition of each ingredient that constitutes the aqua feed should be known. There are two major classifications for feed formulation methods. They are manual method and mathematical or statistical method. Manual method includes Pearson's square Method and Double Pearson's square method. Mathematical or statistical method includes linear programming and algebraic solution using simultaneous equation (Equation 2 and Equation 3). Pearson's square method is used when only two ingredients are used. Double Pearson's method is used when more than two ingredients are used.

Procedure for Pearson's Square Method

1. Per cent of protein needed in the feed ration should be placed in the center of the square.
2. Per cent of crude protein in the basal feed should be placed in the upper left corner.
3. Per cent of crude protein in protein supplement in the lower left corner.
4. Connect the diagonal corners of square with lines.
5. Then, subtract diagonally across the square preferably smaller figure from the larger.
6. Place the answer in the opposite corner.
7. Now, add the figures in the right side corner of the square.
8. Find out relative per cent of inclusion from the total values.

Procedure for Double Pearson's Square Methods

- ☆ Group the ingredients into two broad categories *viz.*, less than 20 per cent crude protein and more than 20 per cent crude protein and take average value.
- ☆ Put the average in the two left corner of the square (Average less than 20 per cent in upper left corner and average of greater than 20 per cent in the lower left corner of the square)
- ☆ Connect the diagonal corners of square with lines.
- ☆ Then, subtract diagonally across the square preferably smaller figure from the larger.
- ☆ Place the answer in the opposite corner.
- ☆ Now, add the figures in the right side corner of the square.
- ☆ Find out relative per cent of inclusion from the total values.

Algebraic Solution Using Simultaneous Equation

Equation (2): A+B = 100 (Two ingredients); Equation of two ingredients is used for determining crude protein (CP) (or) Total Digestible Nitrogen (TDN).

Equation (3): A + B + C = 100 (Three ingredients); Equation of more than two ingredients is used for determining crude protein (CP) and Total Digestible Nitrogen (TDN).

Linear Programming (LP)

LP is to determine which combination will provide the desired nutrient levels at the lowest possible cost. It is a quantitative procedure by which limited resources such as capital, raw material, manpower etc., can be allocated, selected, scheduled or evaluated to achieve an optimal solution to a particular objective. This objective may be to minimize cost or maximize profit. LP was first employed in animal feed industry in mid fifties (to obtain least cost feed). Application of LP in aqua feed formulation is a recent development. Linear – relationships involved in solving the problems; Programming – planning. Therefore, LP involves the planning of activities in order to obtain an optimum result among all possible alternatives.

It is not practically possible for small scale aquaculturist when choices of ingredients are limited. It is a mathematical and statistical technique in which one can develop least cost feed which require the following:

1. Nutrient content and digestible energy of ingredients.
2. Unit price of a feed ingredients including vitamins and minerals.
3. Any other additives (fillers) used.
4. Minimum and maximum per cent of amounts of each ingredient in the feed.
5. Amino acid profile of fish as well as feed.
6. Physical or non-nutritive limitations.

The more information is available on nutritional and physical characteristics of ingredients, the more effective in formulating low cost nutritionally adequate feeds. Common spread sheets such as lotus 1-2-3, MS STAT, INDO STAT are normally used for least cost feed formulation.

Self Assessment Questions

1. What is diet formulation?
2. What is the importance of diet formulation in aquaculture industry?
3. What are food constituents?
4. List out the factors considered in diet formulation.
5. Classify the different methods available for diet formulation.
6. Compare single and double square methods.
7. What is LP?
8. Equation (2): A + B = 100 and Equation (3): A + B + C = 100 meant for what?
9. Important points to be entered in LP technique.
10. Is it LP technique is suitable small scale farmers.
11. Mention spreadsheets available for least cost formulation.
12. Work out the following problems:

Problem 1: Calculate the required percentage of individual ingredients that contain 30 per cent protein by using following ingredients.

1) GNOC – 45 per cent; 2) Rice bran – 12 per cent

Problem 2: Calculate the required percentage of individual ingredients that contain 35 per cent protein by using following ingredients.

Tapiaco flour: 2 per cent crude protein

Wheat brain: 12 per cent crude protein

Fish meal: 60 per cent crude protein

Prawn head waste: 35 per cent crude protein

GNOC: 45 per cent crude protein

Squilla meal: 45 per cent crude protein

Problem 3: Compute a concentrated mixture of 32 per cent protein with the help of simultaneous equation from these ingredients.

Rice brain 9 per cent; GNOC 31 per cent

Problem 4: Calculate a conc. mixture of 18 per cent crude protein and 70 per cent total digestible nitrogen from the following ingredients. (Use simultaneous equation method)

Ingredients	*Crude protein (per cent)*	*Total digestible N_2 (per cent)*
GNOC	40	70
Maize	9	76.5
Wheat bran	14	61

Problem 5: Calculate the required percentage of individual ingredient that contains 2500 kcal/kg by using wheat bran (1663 kcal/kg) and fish meal (4371 kcal/kg)

Chapter 11
Micro Particulate Feeds

It is otherwise called as micro particulate diet. It is an artificial feed and substitute diet for live feed. It is mostly used as fry, grow out, aquarium feeds etc. Micro feeds are broadly classified into wet micro feed or wet micro particulate diet (egg custard) and dry micro feed or dry micro particulate diet. Dry micro feeds are further classified into three major categories, namely, Micro Encapsulated Diet (MED), Micro Bound Diet (MBD) and Micro Coated Diet (MCD).

Wet Micro Feed or Wet Micro Particulate Diet (Egg Custard)

It is a substitute diet for live feed. It is an important diet used in fish hatcheries. Application of egg custard is done generally from PL 8 onwards in shrimp hatcheries. Scheduled feeding rate is as follows:

PL 8 to PL 10 – 15g/feed × 2 times

PL 11 to PL 15 – 25g/feed × 3 times

PL 16 to PL 20 – 40g/feed × 5 times

However, feeding rate with egg custard is done on demand and care should be taken to siphon out the left out feed daily. A broad out line of the preparation of egg custard is given below:

Ingredients Required

Eggs – 36 numbers; Cod liver oil – 75 ml; Yeast – 75 ml; Beef liver – 130 g; Polychaete worms – 150 g; Vitamin mixtures – 20 g; Flesh of squid or prawn – 200 g; Milk powder – 75 g.

Outline Procedure

Mix all the ingredients in to a custard form by using a mixer

↓

Cook it in a pressure cooker for about an hour

↓

Take required quantity of the cooked material
and sieve it through screen
(500 µ) which gives fine granules

↓

Wash the granules till all the fat content is removed

↓

Squeeze out the water and mix it with water
and broadcast in to PL tank

↓

Remaining part can be stored in a refrigerator for further use.

Dry Micro Feed or Dry Micro Particulate Diet

Micro-Encapsulated Diet (MED)

MED are made by encapsulating a solution or suspension of diet ingredients with a membrane. The size of micro-encapsulated feeds ranging from 5 – 700 µ. An microcapsule is characterized by a distinct wall surrounding core material. An ideal microcapsule wall should be impermeable to nutrient linkage and remain intact until fish consumes the microcapsule. Once the microcapsule is eaten, then its wall should be readily broken and liberating the capsule content for digestion and assimilation. The examples are given below:

Dry MED: Nylon protein, gelatin – acacia capsule;

Wet MED – Micro-encapsulated egg diet (MEED)

Advantages

- ☆ Reduce the leakage of nutrients into the surrounding medium *i.e.* water; preventing elevated bacterial concentration and possible out break of disease.
- ☆ Used to deliver nutrients to suspension feeders such as larvae of fin, shellfishes, molluscs etc.

Disadvantages

- ☆ Since emulsification process involved in preparation, it reduces dietary lipids.
- ☆ Capsule walls are permeable to low molecular weight materials such as Amino acid and water soluble vitamins (lipid walled capsules, liposomes).

Dry MED–Nylon Protein MED

An out line of procedure for nylon protein MED is shown below:

Cyclohexane (25 ml) + Span 85 (0.05 ml)

↓

Diaminohexane solution (0.5 ml)

↓

Diet ingredient solution (2.5 ml)

↓

Emulsify for 3 minutes using a homogeniser

↓

Cyclohexane (10 ml), Sebacoyl chloride (0.2ml),
Cyclohexane (30 ml)

↓

Precipitate (MED)

↓

Wash with cyclohexane (100ml) 2-3 times using homogeniser

↓

Precipitate

↓

Sucrose monolaurate (7ml)

↓

Stir it for 24 hrs

↓

Wash for 24 hours in water (2 litre)

↓

Filter with sack cloth of 77 μ mesh

↓

Wash with running water

↓

Nylon protein MED

Wet MED – Micro-Encapsulated Egg Diet (MEED)

Micro–encapsulated egg diet is the typical example for wet micro encapsulated diet. The merits of this feed are easily acceptable to larvae, easily digestible, high nutritional value, low BOD, High water stability, Ease of operation, Wholesomeness, Minimize the loss of nutrients in the environment, Minimize organic load and minimize alteration of O_2 and pH. Finally, it is immense use for researchers to rear tiny PL to juvenile stage for experimental works. Equipments and materials required are centrifuge, test tube, beaker, mixture and glass rod.

Ingredients and their Composition (g/100 g)

Egg	-	97 gm
Glycine	-	1 gm
Vitamins mixture	-	1 gm
Mineral mixture	-	1 gm

An out line of preparation of Micro-Encapsulated Egg Diet (MEED) is given below:

Deshell the whole egg and transfer the whole content into the beaker.

↓

Add the premix (vitamin and mineral mixture) and add glycine which serves as a chaemoattractant and stimulation

↓

Mix the content for 1 minute.

↓

Add hot water to the egg slurry with constant stirring using glass rod

↓

Centrifuge this mixture at 5000 rpm in 15 minutes and collect the supernatant liquid.

↓

Separate the precipitate and store it in 4°C for further use

Feeding Rate

Post larval stage up to PL_{45} – thrice a day.

PL_{40}–PL_{70} – twice a day along with clam meat

Micro Bound Diet (MBD)

In micro bound diet (MBD), the powdered particles are held together by a binder. The dietary ingredients are mixed together with the binder in slurry's and slurry is dried by freeze drying. The binders used are agar, gelatin, carageenan and algin. The size of micro bound diet ranging from 50 to 700 µ. The MBD is typically fed to larval stages of fish and invertebrates. The characteristics of these feed are:

1. The particles are held together by a binder, which may be a complex carbohydrate or protein having adsorptive and adhesive properties.
2. The diet differs from micro encapsulated diet, which has a distinct wall or capsule, surrounding a central core of material.
3. Palatability, nutrient stability and availability and particle stability.

An out line of procedure for MBD preparation is given below:

Ingredients Required

Shrimp flesh, fish flesh, milk powder, egg yolk, soybean oil, vitamin mixtures, binder (wheat gluten).

Take ingredients except vitamin mixtures

↓

Mix with water

↓

Heat in water bath at 80°C

↓

Mix well with homogeniser

↓

Cool

↓

Add binder and vitamin mixtures

↓

Freeze dry / vacuum dry / oven dry

↓

Crush and sieve

Classification

According to production process, it is classified into 3 categories

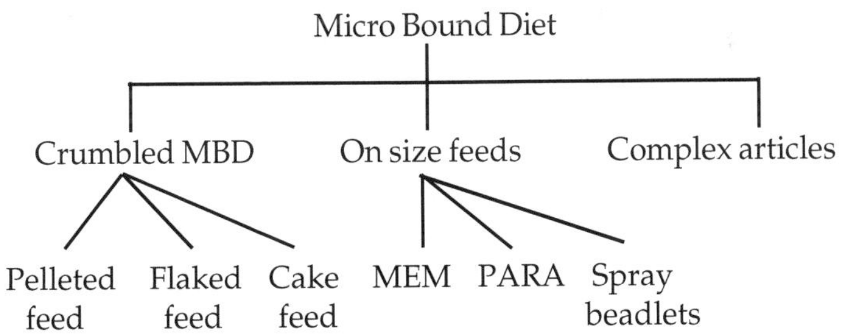

I) Crumbled MBD

It is produced by manufacturing a pellet or flake or cake form.

a) Crumbled MBD–Pelleted Feed

Pellets used in the manufacture of crumbled feeds can be produced through steam pelleting, cooking extrusion or cold extrusion.

Steam pelleting feeds refers to a process by which solid pellets are formed by forcing a feed mixture through holes in a rotating die after pre-conditioning with steam to a temperature of 70–80°C and moisture content of 7–13 per cent. As the pellets emerge from the die, they are cut to desired length by an adjustable knife. The steam partially gelatinizes dietary starch, which aid in binding the ingredients. Precooked corn – sorghum, potato and palm nut, wheat and tapioca starches are sometimes added to the diet at 10–20 per cent and hydrocolloids such as lignin sulfonate and Carboxyl Methyl Cellulose (CMC) at 0.25–5.0 per cent

Cooking extrusion is a process by which a feed mash is moistured with 20–25 per cent water, precooked at 100–150°C and then forced through a die.

Cold extrusion is the process of processing a wet mash (more than 25 per cent moisture) through holes in plate without addition of heat, producing a needle that is broken to form pellets. This type of extrusion can generate pellets with diameters as small as 0.1 – 0.5mm. (disadvantage – moist and semi moist feeds can only be produced).

Crumbled MBD–Flaked Feeds

For a long time, flaked feeds have been the most common type of feed fed to aquarium fish. Although a variety of methods can be used to produce flakes, the double drum dried affords the greatest control of variables affecting flake quality. The equipment consists of two-parallel drums rotating in opposite directions that are heated internally with steam. Feed ingredients are ground to approximate 0.1mm is blended with water to form a slurry or dough that is coated into the drum surface. The dough is flattened to a uniform thickness between the rotating drums and dried to a thin sheet. A blade open the sheet from the (circumference) drum at a point approximately 2–3 drum circumference from the rip of the rolls. The thickness of the flake can be adjusted by changing the distance between the drums. The dried sheet is then crumbled to produce flakes and is screened to produce small particles. The resulting feed has a high surface area to volume ratio and will float for a long time before saturating with water and sinking.

Disadvantages

In flaked feeds high temperature required for drying, proteins can be burned; lipids oxidized, vitamins get lost by exposure to high temperature during any manufacturing process.

Crumbled MBD–Cake Feeds

Crumbled cake refers to feed manufacturing process, in which, a mixture of feed ingredients and minding agents are gelled into a matrix that is dried. Then, it is crushed and sieved into appropriate size particles. Many different binders are used to produce cake feeds including agar, algin, and carrageenan, egg albumin gelatin and zein. The crumbled cake method is currently being used commercially to produce very effective larval feeds.

II) On Size Feeds

a) On size MBD–Micro Extruded Marumerised feeds (MEM)

Microextruded marumerised (MEM) is a two-step process adopted from the pharmaceutical industry to manufacture small, preshaped larval feeds.

First Step

A wet mash of finely ground feed ingredients is first formed into thin noodles by using extruders designed to reduce the operating pressure and twin drum extruders are capable of producing noodles as small as 500 – 300 μm.

Second Step

Noodles are then brokened and shaped in a marumerizers by radial discharge. The higher the speed, the more energy is transferred to the feed causing more shaping and more centrifugal force.

Marumerisers imparts two effects to the feeds break the noodles into length resulting into spheres; to increase the surface density of the feed particles.

Many types of binders can be used with MEM particles provided they are moisture and pressure activated. No heat is added during this process. But, some heat is generated at the extrusion screen due to friction. Feed produced by MEM can be characterized as smooth and spheroid with high density. The smooth shape may decrease nutrient leaching by decreasing the surface area to volume ratio, relative to a rough particle.

b) On Size MBD–Particle Assisted Rotationally Agglomerated Feeds (PARA)

Particle Assisted Rotational agglomeration is a processing method that utilizes a marumerises without extruded noodles. Wet mash is placed directly into the marumerises with a charge of inert particles which in turn transfers energy to the mash producing spheroid particles in a wide range of sizes. An advantage of this process as compared to MEM is lower capital expenditure and lower operating costs due to elimination of the extruder from the process.

MEM	*PARA*
– High capital expenditure	– Lower capital expenditure
– High Operational Cost	– Lower Operational Cost
– Produced with extruded noodles	– Produced without extruded noodles.

c) On size MBD–Spray Beadlets

Spray Beadlets are small micro bound feeds that trap high molecular weight, water soluble nutrients such as starch and protein within gels of alginate or gelatin particles are produced by spraying a slurry of dietary components and a selected gelling agent (*e.g.* alginate) into a curing bath.

III) Complex Particles

Complex particles are produced by combining 2 more techniques (including micro encapsulation) to exploit the advantages and overcome the advantages of individual production methods. This is an exciting development in larval feed manufacturing that allows for production of feeds that more closely for specific needs. For example. Microcapsules or crumbled cake particles may be embedded within MEM, PARA or spray beadlet particles. The complex particle combines the advantages of several types of feeds.

Micro Coated Diet (MCD)

The MCD are made by coating MBD with a coating of cholesterol (or) lecithin (or) Corn gluten or Zein.

MED	*MBD/BCD*
– It has distinct wall	– No distinct wall
– Ingredients not binded	– Ingredients bound
– Size (5-700)	– Size (50-700)
– Used for larval feed	–Used as larval, grow out, aquarium feed
– Examples (Nylon protein MED)	– Examples (Crumbled, MEM, PARA etc.)

Self Assessment Questions

1. What is the other name given for micro feed?
2. Define the term micro feeds
3. Classify the micro feeds according to production.
4. Merits and demerits of MED over MBD/MCD.
5. Compare PARA vs MEM
6. Compare MED over MBD/MCD.
7. What is MEED?
8. How flake feeds are produced?
9. What is egg custard and how it is prepared?
10. Advantages and disadvantages of using MEED and how MEED is prepared?
11. Collect the different brand names available in the aqua clinic for MED and their rates?

Chapter 12
Feed Ingredients

It is a constituent materials or raw materials and also one of the main elements used for the preparation of fish feeds. Feed ingredient nomenclature and classification – (International Feed Vocabulary–IFV) accepted worldwide. IFV has designed a comprehensive name and number to each ingredient using descriptions from one or more of six categories. The categories are origin, which includes scientific and common names for plants and animals; Part fed to animal as affected by processing; Processes and treatments to which the feed ingredients were subjected; Stage of maturity and development; Cutting; and Grade.

Over 18000 feed ingredients have been assigned numbers using this system. Animal and fish feed ingredients are for the most part by products from the human food processing industry. A wide variety of ingredients are available for the use in fish and crustacean feeds.

New (1987) recognized 10 such categories, namely (1) Grasses. (2) Legumes, (3) Miscellaneous fodder plants, (4) Fruits and vegetables, (5) Root crops, (6) Cereals, (7) Oil bearing seeds and oil cakes, (8) Animal products, (9) Miscellaneous feed stuffs, (10) Additives.

Probably, Hardy (1980) recognized 8 such categories namely, (1) Dry forages and roughages (with 18 per cent crude fibre);

(2) Pasture or fodders; (3) Silages; (4) Energy feeds; (5) Protein feeds; (6) Mineral supplements; (7) Vitamins and (8) Additives.

Feed Ingredient Classification

Two major types namely, Nutritive feed ingredient and Non nutritive feed ingredient (additives).

Nutritive Feed Ingredient

Nutritive feed ingredients are of two types, namely, Major and Minor nutritive feed ingredients. Major nutritive feed ingredient includes conventional based, plant based feed ingredient as well as animal based feed ingredients. Minor nutritive feed ingredient also includes non-conventional based, plant based feed ingredient as well as animal based feed ingredients.

Protein Supplements (Non-Energy Feeds)

They are feed ingredients having a protein content of above 20 per cent on as fed or wet weight basis. There are three general groups under protein supplements. They are:

I Group

It is made up of ingredients having a protein content of 20–30 per cent which contains materials of plant origin.

II Group

It is composed ingredients having a protein content of 31–50 per cent and it includes oil seed meals, crab meal and dried milk products.

III Group

It contains ingredients of over 50 per cent protein and includes fish meal, blood meal, feather meal, meat, yeast products, shrimp meal, poultry by products meal, corn gluten meal and casein.

Basal Feeds (Energy Feeds)

They are low protein, high energy feed ingredients. The upper limit for protein content of basal feed is 20 per cent although most are in the 10–17 per cent range.

Plant Based Source Major Nutritive Feed Ingredient

(A) Grasses

Grasses are of high fibre content. Normally, it is utilized either fresh *i.e.* green fodder (as pasture) or in the form of silage or hay. Dried grass is also used in feeds for other livestock also. Dried grass is a potential minor ingredient in fish and shrimp feeds as a source of carotenoids. Since, it contains high fibre content, grasses are of limited value in fish feeds except for herbivorous fish.

Dried grass: If available, this ingredient can be used as a source of protein and vitamins. It contains vitamins of the B group and more importantly, ***b***-carotene, a precursor of vitamin A.

Its high fibre content and its cost would prevent more than a minimal inclusion rate. If added as a source of vitamin A, its potency must be measured because the amount of ***b***-carotene in grass meals depends on standards of drying, the grass species and variety, age at harvest and age of the product. With the limitations noted above, undried grass could, if well chopped, be included in moist rations.

Green fodder includes aquatic and terrestrial plants, mainly as feeds for Grass carp and Breams and some as feeds for Common carp, Crucian carp and Tilapia. There are numerous varieties. The main aquatic plants are *Wolffia arrhiza, Lemna minor, Vallisneria spiralis, Potamogeton malainus, Potamogeton maackianus, Hydrilla verticillata, Eichhornia crassipes, Pistia stratiotes* and *Alternanthera philixeroides*. The terrestrial plants are *Echinochloa crusgalli, Pennisetum alopecuroides, Lolium pereme and Sorghum sudanense, Pennisetum purpurcum* of the grass family; *Lactuca tenticulata* of the composite family; leaves and vines of melon crop and vegetable crop. If water letture (*Pistia stratiotes*), water hyacinth (*Eichhornia crassipes*) and water peanut (*Alternanthera philoxeroides*) are directly applied, fish do not eat, but after proper processing *e.g.* being minced into pieces or fermented, they are preferred by fish. If 100 kg of the above mentioned aquatic plants mixed with 3–4 kg of rice bran and 0.5 kg yeast are sealed for fermentation that can be brought out for use after 2 days at 26°C. Water peanuts containing toxic saponin are not favoured by fish. If they are processed by adding a little table salt with a 2–5 per cent concentration to eliminate the toxicity, fish will eat. If these plants are mashed into grass paste with a high-speed masher, these aquatic plants are desirable in fry culture; since fry can swallow the mesophyll cells in paste, which are similar in size to zooplankton and phytoplankton. The left-over serves as manure for the reproduction of plankton and other natural food.

This kind of green fodder contains plenty of moisture and cellulose. It has little other principal nutrients; carbohydrate, but rich Vitamins such as Vitamins C, E, K etc. It is principal feed for Grass carp and Wuchang fish and serves as supplemental feeds for the other cultivated fish. The proximate composition of nutrients in different green fodders is listed below:

Nutrients of Green Fodder (per cent)

Fodder	Moisture	Crude Protein	Crude Fat	Crude Fiber	Non-Nitrogenous Extract
Wolffia arrhiza	96.02	1.25	0.41	0.38	1.52
Lemma minor	95.80	1.43	0.38	0.42	1.23
Vallisneria spiralis	96.77	0.61	0.09	0.66	1.17
Potamogeton malainus	89.41	2.11	0.17	2.17	4.89
Potamogeton maackianus	87.18	2.16	0.46	3.11	5.65
Leersia japonica	74.61	3.72	1.27	7.50	10.70
Sorghum Sudanense	82.83	1.78	0.69	4.75	8.66
Lactuca tenticulata	88.95	3.02	0.95	1.60	4.20
Pistia stratiotes	91.90	1.20	0.40	1.80	2.90
Eichhornia crassipes	94.90	1.0	0.20	0.90	1.80
Alternanthera philoxeroides	77.5	3.22	0.8	2.62	11.92
Lolium pereme	85.35	4.17	0.59	3.54	4.43

(B) Legumes

Usually, legumes are high in protein content particularly in foliage and seeds. The leaves and stems of legumes are, like the grasses, widely used as fodder for terrestrial animals. A few (*e.g.*, ipil-ipil and alfalfa) have successfully been used in feeds for aquaculture. Legume fodder is rich in protein and minerals. The seeds of legumes have a great potential value as aquaculture feed ingredients though many contain anti-nutritive factors when raw; processing (heat treatment) usually renders them safe for use.

Leguminous seeds are often rich in lysine though poor in methionine. Whole beans and peas are used extensively as human food. They are therefore, usually expensive ingredients and their use in aquaculture feeds may only be justified in diets for high-value export-oriented species. Some examples of leguminous plants, which all have the ability to convert gaseous nitrogen into protein, are acacia, clover, lucerne, groundnut (peanut), gram, lentil, locust beans, chickpea, guar, ipil-ipil, lima beans, field peas, mung bean, cowpeas, and soybean. Some leguminous plants produce high-oil seeds, which are processed for the extraction of vegetable oil.

Legumes are potentially a valuable feed resource for aquaculture in the tropics because of their widespread distribution in those areas. As feeds, they are rich in protein and minerals. Despite the many varieties that have been identified for use as fodder for livestock, only a few have been utilized commercially for feed manufacture. The raw seeds of many legumes contain anti-nutritive factors. This problem can usually be overcome by processing. Most varieties have not been evaluated as feed for fish.

Foliage of several legumes has been utilized in aquaculture feeds. Parts of other leguminous plants is probably also satisfactory, but its successful use has not yet been specifically reported in the literature. Amongst those plants, whose foliage has been used are alfalfa (lucerne) and ipil-ipil (wild tamarind).

Ipil-ipil is a deep rooted tree or arborescent shrub cultivated widely as a fodder crop. Its leaves and seeds contain the glucoside mimosine which is reduced in level if they are stored for a week before use or are soaked in water and dried. It has been used successfully in shrimp and fish feeds at low levels (5-10 per cent). Alfalfa is a deep rooted perennial herb extensively grown for fodder purposes and as a major compound feed ingredient. It is a safe and valuable ingredient for fish feed, contributing protein and fat-soluble vitamins. In compound feed, it is included as a dried meal. Like the grasses, the leaves and feed of legumes, are widely used as fodder for terrestrial animals. Leguminous seeds are often rich in lysine but poor in methionine. *e.g.* For incorporating into fish diets are acacia, clover, lateral, peanut, gram, cow peas, ipil-ipil, soybean, etc.

Red Gram or Dhal (*Cajunus cajan*)

Red gram is grown primarily as food for humans. It is one of the most common legumes of the tropics and sub tropics. The upper (leafy) part of the plant has high protein content and is usually fed to cattle. The seeds are often used up to 30 per cent in poultry rations. Red gram bran, a by-product of gram flour production, has almost the same protein content as the seed and is used for the same purpose. Although not evaluated as feeds for fish, both seeds and bran are judged safe for incorporation into fish feeds at moderate levels.

Chick Pea, Bengal Gram (*Cicer arietinum*)

The chick pea is native of India but is also cultivated in other parts of the tropics. Although grown mainly for human consumption, the seeds of the chick pea are frequently used as feed for poultry and livestock. Although not evaluated as a feed for fish, it is presumed to be safe for inclusion as a major constituent in compound fish feed. The leaves of the plant are also suitable for feeding to fish.

Lablab or Egyptian Bean (*Dolichos lablab*)

A perennial plant native of India, the lablab is grown in many parts of the tropics. The seeds of the lablab, which are fairly rich in protein, have been fed as concentrates to livestock, but have not been evaluated in fish. This plant is not to be confused with lab-lab, an algal complex fed to milkfish in the Philippines.

Lentil or Red Dhal (*Lens culinaris*)

The red dahl is native of Asia, but is also grown in Africa. The plant is cultivated for its seeds which are used primarily for human consumption. During the milling of the seed, bran rich in protein is produced. The material has a high feed value and, although not evaluated in fish, probably can be used at fairly high levels in compound fish feed.

Ipil-ipil or Kathin (*Leucaena leucocephala*)

The ipil-ipil is one of the two legumes whose leaves and stems have been used commercially on a regular basis in animal feed manufacture in many parts of Asia. Ipil-ipil leaf meal is fairly rich in protein but may not be fed at excessive levels to farm animals due to the presence of the toxic amino acid mimosine. Ipil-ipil has been used at low concentrations (5-10 per cent) in compound fish feed in Thailand and the Philippines without noticeable adverse effects. One study conducted on tilapia showed very poor growth when

Leucaena constituted 25 per cent or more of the dietary protein (Jackson, A.J., B.S. Copper and A.J. Mathy, 1982).

Lupin (*Lupinus* spp.)

The lupins are grown in the cooler parts of the tropics. Both the leaves and seeds are rich in protein; the latter fed up to 20 per cent in rations for monogastric animals. Although not evaluated as a feed for fish, it is believed that the seeds of the lupin can be fed at moderate amounts to fish without harm.

Lucerne or Alfalfa (*Medicago sativa*)

The alfalfa is the better known of the two legumes whose leaves and stems are exploited commercially as a principal ingredient in compound feed. Alfalfa meal, unlike ipil-ipil, does not appear to contain anti-nutritive factors that limit its use in feeds. The protein content of alfalfa is comparable to that of ipil-ipil. Alfalfa is known to be safe for incorporation into compound fish feed. Crude protein: 13 – 17 per cent; Crude fibre: 25 – 30 per cent; Good source of minerals (Ca, K, Fe, Mn and Zn); Good source of vitamins (choline, biotin, niacin, pantotemic acid, riboflavin, tocopherol).

Velvet Mesquite (*Prosopis veleutina*)

A shrub native of Mexico, the mesquite produces an abundant crop of pods which are sugary and contain seeds rich in protein. The seeds are encased in a hard shell which must be crushed to enhance digestibility.

Saman or Cow Tamarind (*Samanea saman*)

The saman is a tree that grows in the tropics. Its leaves are fairly rich in protein but also high in tannins. Its seeds are also used as concentrates for livestock. Both leaves and seeds have not been properly evaluated in fish although preliminary studies in Venezuela have shown that the ground pods fed at moderate levels have no adverse effect on the cachama (a native species belonging to the Characidae family).

Sesbania (*Sesbania grandiflora*)

These small fast-growing shrubs are very common in the tropical lowlands of Asia. Its leaves are rich in digestible protein and its seeds are often fed as concentrates to poultry. Although not evaluated in fish, ready acceptance of both leaves and seeds by poultry indicates its safeness.

Mung Bean or Black Gram (*Phaseolus mungo*)

The mung bean is a native of central Asia, grown primarily for the bean which serves as food for humans. Both leaves and seeds are fairly digestible for monogastric animals. Although not evaluated as feed for fish, the absence of glucosides in the seeds makes them safe for use at moderate levels.

Red Bean or Rice Bean (*Vigna umbellata*)

Grown primarily for its seeds which are used for human consumption, the red bean is found throughout tropical Asia. It has not been evaluated as a feed for fish, but only the seeds appear suitable for use in compound feed.

Horse Gram (*Dolichos biflorus*)

Also an herb found in tropical Asia; the seeds of the horse gram are used as concentrate feed for cattle. Although not evaluated for fish, because of their low fibre content, these seeds appear suitable for use at moderate levels in compound fish feeds.

Cow Pea or Black-eye Pea (*Vigna unguiculata*)

The cow pea is grown primarily for human consumption. Its seed is usually too expensive for use as feed for livestock. Its leaves, however, are equally good and appear quite suitable for use in fish feeds, although no evaluation has yet been made.

Leguminous Seeds with Aquaculture Feed Potential

Common Name	*Characteristics*
Horse gram	Low fibre
Cow pea	Low fibre
Mung bean (black gram)	No glucosides; successfully used in aquaculture feeds
Tamarind	High fibre
Sesbania	
Saman (rain tree)	High protein; high in tannins; used in experimental fish feeds
Red gram (dhal)	As good as soybean for poultry
Split peas (red dhal)	High protein bran
Lupin	High protein; bitter varieties contain alkaloids and must be soaked before use
Algaroba	Ripe pods palatable to other animals
Senegal gum	High protein

Most of the leguminous seeds eaten by humans may be expensive for use in aquaculture feeds, but those of other legumes, either alone or together with the foliage of the plant, remain relatively unexploited sources for aquaculture. However, care should be taken to check their palatability and to remove toxic substances before their large-scale use.

(C) Miscellaneous Fodder Plants

The leaves and other aerial parts of many plants, other than those specifically grown for fodder are used for this purpose. While these may have local significance as aquaculture feed ingredients, nutrient digestibility (though crude protein levels on a dry matter basis are often quite high) is low. Sisal hemp leaf waste pulp could be used in small quantities in moist feeds, but its protein is not very digestible. The leaves of the papaya, which are high in protein yet quite low in fibre could be a useful ingredient. The leaves of coffee plant are low in protein but high in fibre. Water hyacinths could be a useful ingredient for moist feeds if first boiled to form a paste. Swamp cabbage, is another water plant, a rooted one, which has similar potential. Small quantities of sago palm starch can be used in feeds–its ease of gelatinization improves feed stability.

These plant parts are pretreated by soaking in water and or sun drying and used as a leaf meal. These plants have been investigated to substitute for fish meal in aquaculture diets. Ng and Wee (1989) reported that Cassava leaf meal could be used to replace up to 20 per cent if fish meal protein. Wee and Wang (1989) used Ipil – ipil leaf meal to replace 20 per cent fish meal protein. The morning glory (*Ipomea* sp.) is used as a fodder for silver barb in Thailand.

Although, there are many varieties of fodder plants which can be fed to fish, their exploitation for this purpose, for the most part, has been hindered by low digestibility of nutrients and difficulty in obtaining adequate and regular supplies. Two exceptions in this category are the sago and the swamp cabbage which are relatively abundant in the tropics and which have high feed value for animals because of their high digestibility.

Kangkong or Swamp Cabbage (*Ipomoea aquatica*)

Mainly cultivated for food, the kangkong has also been used as stock feed in South East Asia. It has a fairly high feed value, with higher protein content than alfalfa. Kangkong has been fed successfully to carps in Sri Lanka as a principal component of compound feed.

Sago Palm (*Metroxylon sagu*)

The sago palm, which often proliferates in freshwater swamps, is valued for the starch contained in its trunk. The starch, which is normally extracted in small village mills, has a variety of industrial uses apart from its value as food. The unrefined starch, or meal, has been fed to livestock and is considered equally safe for fish. Small quantities incorporated into compound feeds improve pellet quality because of the easy gelatinization of sago starch under the conditions of pelleted feed manufacture.

(D) Fruits and Vegetables

Waste fruits and vegetables and the by-products from their processing or harvesting have not been used much in compound feeds for aquaculture. However, they are sometimes used, fresh, for feeding or manuring ponds. As both types of products are usually seasonal, large quantities of wastes and by-products are available intermittently, but they are not normally preserved for later use by ensiling or drying.

The leaves of these plants are usually more nutritious than the stems. Examples of the many by-products in this category, which are often wasted, are those from grapes, tomatoes, the palms, bananas and plantains, mangos, melons, citrus fruits, pineapple, and breadfruit.

In many countries, large quantities of citrus wastes are available. Whole fruits are sometimes wasted when they cannot be marketed quickly. Dried citrus pulp, a waste from juice production, citrus molasses and citrus seed meal, oilcake are available. However, the seed meal is toxic to chicken and pigs and may be so to fish. Most wastes from fruits *e.g.*, coffee plant are very high in fibre and therefore of limited use in fish feeds. Coffee wastes are also unpalatable to other livestock and depress growth. Banana plants have some potential as ingredients. Banana waste has been used as a binder in crustacean feeds. Banana peels are rich in tannins and cannot be used until they are yellow. Limited quantities of various banana wastes could be utilized as sources of carbohydrate.

(E) Root Crops

Root crops are rich in CHO and an excellent source of energy for many classes of livestock. However, their value as ingredients for aquaculture feeds in limited. This is because of high value for human food and inability of most finfish species to digest CHO's. (Although fish, being poikilotherms (cold blooded), are less capable of utilizing carbohydrates at low temperatures, they utilize fats and proteins at least just as well as land mammals. Fish, the herbivorous species excepted, are not capable of utilizing cellulose). Waste from root crops can be utilized in small quantities in compound feeds, but, generally many require heat treatment to destroy the toxins. Some root crops have special value in aquaculture because of their ability to increase the water stability of diets *e.g.* Potato and cassava starches that one commonly used as binders. Other plants in this category include Yams, Carrots and Jerusalem antichakes. Eg: Tapioca, Sugar beet molasses, meals made from potatoes and sweet potatoes. Presence of hydrocyanic acid in tapioca should be monitored.

For the most part, roots and tubers are grown as food for human consumption in the tropics. However, because of their high starch content, some varieties are grown as industrial raw material. Notably, among these is cassava or tapioca. Roots and tubers are also suitable for feeding to fish, although their use for this purpose appears very limited due to their relatively high cost. Also, because the feed value of roots and tubers is in their carbohydrate content, they cannot be used at too high levels in feed for fish, which have a lower tolerance for carbohydrates than farm animals.

Sweet Potato (*Ipomoea batatas*)

Sweet potato is mainly grown as food for human consumption in the tropics. The tubers consist primarily of carbohydrates with a high content of sucrose. Protein content, although low, is almost twice that of cassava. The leaves of the sweet potato are highly digestible, free from toxins, and fairly rich in protein. The starch in the tuber provides good binding for dry and moist-type pellets if heat processing is employed.

Cassava, Tapioca or Yucca (*Mannihot esculenta*)

Fresh cassava is seldom used as feed for fish. Most varieties are toxic unless properly processed to prevent the release of hydrocyanic acid by an enzyme present in the tuber. Heat applied during processing destroys the enzyme. Cassava has low feed value in diets for fish but its presence in small quantities improves the water stability of pelleted feed. Fresh cassava has been included in the diet of milkfish in Indonesia (Hastings, W., 1975).

(F) Cereals and its by Products

Cereals and cereal by-products, despite their high carbohydrate content, form an important component in aquaculture diets. Their starch content helps to increase the water stability of the feed, particularly where heat is included in their method of processing.

Cereals also contribute significantly to the protein and lipid content of the diet. Though deficient in some amino acids (*e.g.*, lysine), they can be used to balance high-protein animal and vegetable ingredients. Cereals are often one of the cheapest raw materials that can be included in compound feeds for aquaculture. The brans are excellent sources of the B group vitamins. The uses of cereal grains are well known. Ground whole cereals which can be used in fish feeds include the millets (various species), oats, barley, sorghum, milo or dari, wheat, and corn or maize. The quantity of grain used depends on its cost and the limited utility of high carbohydrate ingredients for fish. However, many cereal by-products are available and potentially useful for aquaculture feeds. The major ones are defined as follows:

- ☆ *Broken Rice*: Damaged rice separated out after rough rice has been dehulled and polished. It has the same chemical analysis as polished rice and is often sold as human food.
- ☆ *Rice Hulls*: An extremely high fibre material, not recommended for fish feeds.
- ☆ *Rice Bran*: A good source of B group vitamins, it is the material scoured off before initial rice polishing. It has higher protein content than the original grain and, in the unextracted form, has a high lipid level which is prone to rancidity. Rancid rice bran is much reduced in feed value. The oil is often extracted for human use and the resultant rice bran is favoured, because of its keeping quality, by feed millers. The fibre in extracted rice bran absorbs water and leads to a water unstable pellet.

- *Rice Polishings*: This is the part of the starch endosperm which is removed during polishing, to improve the appearance of the polished rice for human use. It is therefore lower in protein and fibre than rice bran. It has the same keeping quality problems as rice bran.
- *Rice Pollards*: A mixture of rice bran and rice polishing, usually sold as a grade of rice bran.
- *Rice Mill Feed*: A mixture of all the rice milling by-products, typically 60 per cent hulls, 35 per cent bran, and 5 per cent polishing.
- *Rice (General)*: There are many intermediate products between those identified above. The secret of their origin is in their chemical analysis. From the original rough rice (padi) about 50-60 per cent becomes polished rice, 20 per cent hulls, 10 per cent bran, 3 per cent polishing and 1-17 per cent broken rice.
- *Wheat Bran*: Contains most of the vitamins and protein of the wheat grain. It is the fibrous coating of the grain beneath the husk. It does not contain the wheat germ. There is a coarse outer layer and a fine inner layer to the bran area. It is quite high in protein and fibre. Like extracted rice bran, wheat bran can cause water stability problems in fish feeds.
- *Wheat Middlings:* This is a mixture of 'shorts' (itself a mixture of fine bran and feed flour–the outer endosperm layer) and wheat germ. It is sometimes known as Wheat Pollards or Mill Run. It is less fibrous than wheat bran, has a higher feed value and contains more wheat protein (gluten). It is thus better from a water stability point of view.
- *Maize (Corn) Gluten Feed*: A by-product of the wet-milling of maize for the manufacture of corn starch or corn syrup, corn gluten feed is usually 20-30 per cent protein and quite high in fibre. It contains the corn colouring pigment. Like other maize products it is very lysine deficient.
- *Maize (Corn) Gluten Meal*: This corn by-product is much higher in protein and lower in fibre content.

- ☆ *Maize-Germoil Meal Maize-Germoil Cake*: This is a high oil/ low protein by-product of the same industry.

It is grown primarily for human consumption, cereal grains are no less important as feed for efficient animal production. Present day, high energy diets for monogastric farm animals often contain up to 80 per cent cereal grains and their by-products. The use of grains is somewhat restricted in fish diets because of generally lower tolerance for dietary carbohydrates by fish. This effect appears to be more pronounced among cold-water species such as the trout than among warm-water species.

While cereal grains are considered mainly as a source of dietary energy, their by-products represent a fairly rich source of protein and polyunsaturated fatty acids (PUFA).

Cereal grains sometimes constitute a significant percentage of fish feed processed into pellets. Starch present in cereals act as good, natural binders when gelatinized under normal pelleting conditions, giving products that have high water stability. Consequently, cereal grains are indispensable in the manufacture of floating-type pelleted feeds.

Millet (in the tropics, mainly *Eleusine coracana, Paspalum scrobiculatum* and *Pennisetum americanum*) grown primarily for human consumption, millet is only occasionally available as feed for livestock. Its use as a feed for fish has not been documented. Judging from favourable growth response of monogastric farm animals fed on the cereal, millet should, at least, be comparable to maize in feed value. The seed is hard and proper grinding is necessary before use in fish diets. Sorghum and millets Crude protein: 8 – 12 per cent; Poor profile of amino acid, minerals, vitamins; used as energy feed.

Rice (*Oryza sativa*)

Rice is seldom used for animal feeding because of its high cost, although damaged grain and portions considered unfit for human consumption, viz., sweepings from warehouses and mills, are available for that purpose. In a few countries, notably Thailand, where it is the principal grain crop, rice of low commercial value such as broken rice is also used as livestock feed. On the other hand, by-products from mills are more generally available and constitute the most important feed resource in all rice-producing countries.

The by-products include: rice bran, rice polishings and rice mill feed. Rice and its mill by-products are important feed constituents in fish feeds.

Broken rice and Cargo Rice

Rough rice is very seldom fed to fish. Instead, broken rice (polished) and cargo rice (unpolished) are used in the preparation of traditional type fish feeds as well as in compound diet formulations. The rice is first boiled in water before mixing with other ingredients to make the ration. Rice is an excellent energy source with a low, but fairly good quality protein content. No digestibility study has been made with fish, but for mono gastric farm animals, uncooked rice is more digestible than maize. Cargo rice, especially, is rich in the B-vitamins.

Rice Bran

Freshly produced rice bran has high oil content (14 to 18 per cent). The oil is sometimes recovered by solvent extraction because of its high commercial value, especially in countries where demand for cooking oil exceeds supply. The oil present in the bran is rich in polyunsaturated fatty acids which undergo rapid oxidation under normal storage conditions. Rice bran that has turned rancid has markedly reduced feed value. Rice bran has a higher protein content than the grain. It is also fairly high in fibre, thus limiting its use in fish feeds. Often, rice bran is mixed with rice polishings at the mill and sold as rice bran. Adulteration with hulls lowers its feeding value considerably. Good quality pellets can be made from feeds containing unextracted rice bran.

Oil extracted rice bran is almost totally devoid of oil. It is the type of rice bran favoured by feed mills because of its longer shelf life. Its protein content is somewhat higher than fresh rice bran while its energy content is lower. Pelleted feed containing a high proportion of de oiled rice bran are not very water stable. This is because of the water absorption characteristics of fibre, which will be fairly high in such feed. Where available, rice bran has been successfully used in compound fish feeds for both warm as well as cold water species.

Rice bran: Crude protein: 10 – 12 per cent; Crude fibre: 12 – 18 per cent; Lipid: 7 – 12 per cent; Ash: 8 – 12 per cent; Good source of energy and B group vitamins. Deoiled rice bran is better.

Rice Polishings

Rice polishings are produced during the scouring of rice after the bran has been removed. The material, therefore, contains a higher proportion of starch but is lower in protein and fibre content. It is not produced in as large quantities as rice bran, and, in most mills, are often mixed with the latter to give rice pollard. This, however, is usually sold as a grade of rice bran in most countries.

Sorghum (*Sorghum bicolor* and *S. vulgare*)

Sorghum is grown in warm areas where rainfall is inadequate for maize cropping. Its feed value is comparable to maize for most species of livestock. In Mexico, where maize is normally reserved for human consumption, sorghum is used as a substitute in compound feed for trout and carp.

Sorghum contains tannic acid which decreases the availability of methionine in the diet. The amount of tannic acid in sorghum depends not only on the variety of sorghum, but also on the conditions under which the crop is cultivated.

Wheat (*Tritricum aestirum*)

Although primarily a crop of temperate countries, wheat is also grown in parts of the tropics where there is a long period of relatively cool weather. Where it is not grown, it is imported for flour production. Wheat products are, therefore, available in practically every country.

Wheat, often used as an ingredient of compound fish feed in temperate countries is seldom used in the tropics because of its high cost. Damaged grain is, however, frequently available and when fed to fish is at least equivalent to maize in feed value.

Milling of wheat produces three major by-products, two of which are almost exclusively used as feed for livestock. These are: wheat bran and wheat middlings (or fine bran). The third, wheat germ, has higher commercial value as a food item for humans. Occasionally, damaged flour is also available as feed.

Wheat Bran

Wheat bran is the primary coat of the wheat grain. It has fairly high protein content. Because it is also high in fibre, it has a laxative effect when fed at excessive levels to animals. Such an effect, however, will be more difficult to determine in fish. Nevertheless, wheat bran has been successfully fed at fairly high levels to various species of fish without adverse effects on growth. Too much wheat bran in a formulation results in pellets with poor water stability due to the water absorption characteristics of fibre.

Wheat Pollard

Wheat pollard is also known as fine bran. The material is less fibrous than wheat bran and has a higher feed value. Wheat pollard also contains a higher proportion of gluten which is a good natural binder for pelleted feed production.

Wheat flour and bran: Good source of energy; Crude protein: 10 – 14 per cent; Crude Fibre: 12 – 18 per cent; Ash: 6 – 18 per cent; Good source of P, K, Mn, Zn vitamins such as Niacin, Biotin; Wheat gluten meal – 20 to 30 per cent Crude protein; Corn gluten Crude protein: 20 – 30 per cent; Good source of Fe and Zn.; Good source of Niacin and Vitamin E.

Maize (*Zea mays*)

Maize is grown throughout the tropics and is the principal cereal grain employed in the feeding of livestock. Maize has also been fed successfully to fish. Whole grain has been used as a supplementary feed for tilapia and carp. However, maize is usually used in finely ground form as an energy component in compound feeds. While there is little published data on the digestibility of starchy components of cereal grains in general, studies on maize have indicated enhanced utilization after cooking. Normal pelleting processes only partially gelatinize the starch granules whereas total starch gelatinization occurs when the cereal is subject to high temperatures under moist conditions as in extrusion cooking. For channel catfish, such processing methods greatly increase the digestible energy value of maize, indicating the relatively low digestibility of raw starch for that species (Lovell, R.T., 1977).

Wet milling of maize is an industrial process for production of valuable components of the maize kernel. The main products of the process are: corn oil and starch (from which dextrins and glucose are obtained after further processing). The by-products include: maize gluten meal, maize bran and maize germ meal. Often, all three are mixed together and sold as maize gluten feed. The protein content of maize gluten meal is comparable to that of soybean meal. However, maize protein has a very low lysine content and maize gluten meal or maize gluten feed should be used along with other lysine-rich protein sources (animal protein, or soybean meal).

(G) Oil Bearing Seed and its By-products

Many plants are grown specifically for the oil which their seeds or fruits produce, which is utilized for human food and other purposes.

Vast quantities of by-products from the vegetable oil industry are produced and these are the staple ingredients of animal feedstuffs, being high in protein and low in carbohydrate. All are potential ingredients of aquaculture feeds.

Examples of the plants from which products in this category come are the leguminous plants soybean and groundnut, together with mustard, rape seed, sunflower, safflower, coconut, kapok, cotton, oil palm, linseed, poppy, sesame (gingelly) and para rubber.

In considering the use of ingredients from this group, it is essential to understand the terminology used in describing oil seed by products because ingredients with apparently similar names have completely different analytical characteristics. The external coating of some seeds is sometimes but not always completely removed before oil extraction, *e.g.*, in the case of sunflower seed, groundnut, and cotton seed.

The material which remains after oil extraction is referred to in several ways:

Decorticated (sometimes referred to as 'Dec')	Coating removed before oil extraction
De-hulled	Coating removed before oil extraction
Without hulls	Coating removed before oil extraction
(Undecorticated (shortened to 'Undec')	Coating not removed
With hulls	Coating not removed

Some intermediary products between 'Dec' and 'Undec' exist. A few tables of feed composition refer to these as 'with some hulls'. Decorticated products are higher in protein and lower in fibre than undecorticated products.

The other major set of terms applied to this class of feeds refers to the method of oil extraction used, which also has important analytical consequences. These terms are defined as follows:

Expeller (shortened to 'Exp'): Oil removed by mechanical process, either by hydraulic presses or by screw augers. The latter type can be distinguished by looking at the pieces of unground cake, which are not flat. The hydraulic process does not remove as much oil, but, damages the cake less than the screw process which generates a lot of heat.

Extracted (shortened to 'Ext'): Oil removed by a highly efficient chemical process using solvents. Sometimes the words 'solvent extracted' are applied to this residual product.

The characteristics of these products are that expeller residuals are much higher in oil content and lower in protein content than extracted products.

Two other terms are applied to oil seed residues. These are 'cakes' and 'meal'. Normally, if a product is referred to as a 'cake' it means it is an expeller residue. Similarly, a 'meal' normally refers to an extracted product. However, there can be some confusion here because the word 'meal' can also be used to refer to a ground or milled product. So, the words 'groundnut meal' might refer either to groundnut cake which had been ground into a meal or it might refer to extracted groundnut. When in doubt, the chemical analysis is the only criterion to use in determining which product is being offered for sale if it is in ground (meal) form. Some illustrations of the important analytical differences designated by the above terms are given below:

Examples of the Effect of Processing on Analytical Characteristics of Oil-Seed Proteins

Country and Material	*H_2O*	*LIP*	*PROT*	*FIB*	*NFE*	*ASH*
Pakistan						
Dec. Cottonseed	7.3	5.2	36.7	8.8	34.9	7.1
Undec. Cottonseed	6.5	8.9	21.5	24.5	32.5	6.1
USA						
Exp. Groundnut	10.8	7.3	45.1	6.8	24.7	5.3
Ext. Groundnut	8.5	1.2	47.4	13.1	25.3	4.5

The major oil-bearing seeds contribute large quantities of by-products which are used in animal feeds, all of which have potential as aquaculture feed ingredients.

The following is a list of the major plants contributing products to this category of ingredients:

Common Name of Oilcake or Meal	*Characteristics*
Groundnut (Peanut)	Methionine deficient; prone to aflatoxin development; extensively used in carp diets.
Mustard (Rape Seed)	Also used in carp diets but needs great care, as non detoxified meals contains a range of toxins.
Coconut (Poonac)	Prone to rancidity; absorbs water, thus water unstable feeds; low in protein; high in fibre.
Oil Palm	Kernel high in protein; fats saturated;
Soybean	High protein; low lipid; generally good source of EAAs including lysine but methionine deficient; contains a trypsin inhibitor and urease but these are destroyed during processing; with EAA supplementation it is a potential partial replacement for fish meal.
Cotton	Screw press cake contains high level of free gossypol which is toxic; high in fibre; potentially valuable but knowledge of effect of gossypol on fish not adequately known
Sunflower	Richer in methionine and cystine than soybean; lysine deficient; no toxins.
Safflower	Poorer in the EAAs lysine and methionine than sunflower.
Para Rubber	Must be de-toxified to remove prussic acid; suitability probably similar to coconut.
Flax (Linseed)	Like para rubber, contains an enzyme and a glucoside that produces prussic acid but normal processing destroys the enzyme; not known whether its additional toxicity to poultry (which can be eliminated by supplementary vitamin B_6) affects fish or not.
Sesame (Gingelly)	Rich in methionine but deficient in lysine; would be good ingredient in conjunction with soybean; value for Indian carps has been demonstrated.

Oil seeds are generally higher in protein and low in the CHO than cereals. (20 – 50 per cent protein). It also deficient in EAA like lysine and thereonine. Examples: oil cake – by products; let after the extraction of oil from oil seeds.

Oilcakes are by-products of the vegetable oil extraction industry. Although many varieties of seeds and fruits are cultivated primarily for their oil content, the protein-rich residues left after oil removal represent an immense resource upon which the world's production of animal protein for human consumption largely depends.

In the tropics and sub-tropics, soybean, groundnut, and sesame (gingelly) are the principal oil seeds yielding protein-rich oilcake or meal after oil removal. Among oil-bearing fruits, the coconut and the oil palm nut are the most important sources.

Oil extraction from seed or fruit is carried out by two methods: by pressing, or with chemical solvents. The product obtained by pressing is termed oilcake and that by solvent extraction, oil meal.

Oilcake production is carried out by one of two mechanical processes. The first process involves the use of hydraulic presses that press out the oil from oil seed, or previously shredded dried fruit kernels (*e.g.*, copra), wrapped in cloth. The resulting residue is in slab form. The presses used in this process are of relatively simple design and can be fabricated quite easily. They are, however, not very efficient. The second process involves the passage of seeds or fruit kernels along a screw auger confined in a steam-heated conically shaped jacket with the oilcake emerging at the smaller end. This process, known as the screw expeller process, is more efficient in removing oil. However, because of the high temperatures associated with the pressure applied, oilcakes produced by this method may come out burnt with subsequent lowering of feed value (especially with regard to protein digestibility in general and lysine availability in particular). Because of the high value of edible oils relative to oilcake residues in developing countries, the screw expeller process is favoured over the less efficient hydraulic press process for mechanical expression of oil from seeds and fruit. However, in South Asia as well as in South East Asia, with the possible exception of Thailand, hydraulic-press extraction continues to be the principal method employed in the vegetable oil industry.

Oil extraction involving chemical solvents is of fairly recent origin. Solvent extraction plants are technically more sophisticated than the screw expeller mills and their operation is economical only for processing large quantities of oil seeds or fruits. These plants are found in most of the countries that came under this survey, although they are more prominent in some than in others. Where they are located, these plants are usually used for extraction of oil from soybeans that are either grown domestically, or imported. The process is now extensively used for secondary recovery of oil in the palm oil industry in Malaysia as well.

The oil meal that is produced as a by-product of the solvent extraction process has a very low fat content; often less than 1 per cent. Temperatures involved in the solvent extraction process are generally low compared to the expeller process. Because of this and because of the need to de-activate antinutritive factors generally present in some oil seeds, toasting of the extracted residues follows solvent recovery. Due to the relatively mild treatment given to oil meals, these products have a comparatively higher feed value than expeller cakes. The lower residual oil content in an oil meal also gives a higher protein content compared to a similar product obtained by the screw expeller process.

Groundnut (*Arachis hypogaea*)

Groundnut oilcake is a safe feed for fish. It has been demonstrated in India that diets consisting of 50 per cent of the material as the principal protein source can be used in complete feeding of carps. Its high polyunsaturated fatty acid (PUFA) content also makes further addition of fats to such diets unnecessary. Because groundnut protein is especially low in methionine, the oilcake should be used with methionine-rich protein supplements or with synthetic methionine to achieve a proper balance of essential amino acids in the diet.

Quality in groundnut oilcake depends on whether the material is made from decorticated nuts. Removal of the fibrous hulls yields a better quality product with higher protein content. This will also depend on the amount of residual oil in the cake. In a few countries, there are two varieties of groundnut oilcake: the mill produced and the 'country' produced. The former is an industrial product with a generally lower residual oil content (less than 7 per cent), whereas the latter is produced on often a very small scale in villages and may contain as much as 13 per cent oil.

The principal (and often very serious) contaminant of groundnut cake is aflatoxin. Aflatoxin is a group of highly toxic substances produced by the mould, *Aspergillus flavus*. The most prominent of these are Aflatoxins B_1, B_2 and G_1. The toxins are produced only when the mould exists as a pure culture. Aflatoxin is a hepatotoxin and mortality among afflicted animals and fish invariably results from severe liver damage. Small doses over an extended period of time produce cancer of the liver in humans. The half lethal dosage, or LD_{50} 1/ for trout weighing 100 g is 0.5 ppm in the diet. Improper post harvest handling of nuts is usually blamed for the presence of aflatoxin in groundnut although pre-harvest infestation has also been documented. Aflatoxin is not destroyed by heat.

1/: LD_{50}, the dosage at which mortality among animals consuming the contaminated feed is 50 per cent.

Coconut (*Cocos nucifera*)

Coconut oilcake is widely available as a feed for fish and livestock in countries such as Sri Lanka, Indonesia and the Philippines. On the average, 1 000 nuts yield 60 kg oilcake (called poonac in South India and Sri Lanka). Compared with residues of oil seed extraction, coconut oilcake is low in protein and high in fibre. It is difficult to make a water-stable pelleted feed if the formulation contains too much coconut oilcake, as the material tends to absorb large amounts of water, causing the pellet to disintegrate. Poonac is commonly used as a supplementary feed in the pond culture of carp in Sri Lanka. It has also been included in the diet of milkfish in Indonesia (Hastings, W., 1975).

One of the main problems in its use is the tendency of coconut fat to turn rancid in storage. The effects of feeding rancid coconut oilcake to fish have not been studied, although short-chain free fatty acids in the diet have been shown to suppress growth in at least two species, the trout and channel catfish. However, rancid soybean oil does not appear to affect the growth of carp.

African Oil Palm (*Elaceis guineensis*)

The African oil palm has been successfully transplanted in South East Asia. In Malaysia especially, the oil palm has emerged as the major source of cooking oil. The residue of the palm kernel has slightly lower protein content than coconut oilcake or meal, but its feed value is not very different. Like coconut oil, palm kernel oil is low in PUFAs. Use of palm kernel oilcake in fish diets has not been documented although, like coconut oil cake, it should be used sparingly in diets for fish.

Soybean (*Glycine max*)

Soybean is one of the most important oil seed crops in the world. A legume, like the groundnut, soybean has relatively low oil content. Because of this the seeds can be easily flaked by crushing, oil removal of flaked soybean is carried out by solvent extraction. The residue is very low in fat (less than one per cent) and high in protein. Decorticated soybean meal may contain up to 55 per cent protein.

Soybean meal is a good source of essential amino acids (EAA) and is one of very few plant sources rich in lysine. In recent years, soybean meal has been increasingly used as a substitute for more expensive fish meal in compound fish feeds. Soybean meal has a relatively low methionine content which can be corrected by addition of synthetic forms of this amino acid in the feed. Anti nutritive factors, principally urease and trypsin inhibitor present in the raw bean are usually destroyed by heat during the oil extraction process.

Full fat soybean processed under high pressure has recently been successfully used as a complete substitute for fish meal in the diets of both warm-water (catfish) as well as cold-water (trout) species. The high PUFA content in full-fat soybean makes supplementing with extraneous PUFAs unnecessary.

Considerable work has been carried out on the extent to which processed soybean products can replace fish meal as a protein source in fish diets. While such replacement can be total for carp without effect on growth similar replacement in diets for trout resulted in slower growth and poorer feed conversion. Soybean oil cake (Dehulled meal in recommended). Among the major plant protein sources, it is considered as best protein source. Carps, tilapia, channel catfish – as high as 50 per cent Seabass, grouper, -10 – 20 per cent; Prawns – up to 40 per cent.

Cotton (*Gossypium* spp.)

Cottonseed *per se* is rarely used as feed. The seed is processed by first removing the lint and then cutting the tough hull to release the kernel. These operations are carried out by machines. The kernels are then crushed and the oil removed by mechanical means or by solvent extraction. In developing countries, the former process is most common. Although cottonseed cake has high protein content, it is rather lacking in lysine. For fish, therefore, it should be used in combination with other materials such as soybean meal and animal protein sources.

An important factor limiting usage of cottonseed cake or meal in fish feed is its gossypol content. Gossypol is a toxic phenolic compound confined to the genus of cotton, *Gossypium*. It is the yellow pigment which constitutes 20 to 40 per cent of the substances inside the glands of the seed kernel. The amount of gossypol is proportional to the number of glands present as it does not occur elsewhere in the seed. Cottonseed usually contains 0.4 to 1.7 per cent gossypol. There have been few studies on the toxic effects of gossypol in fish although from animal studies, it has been shown to inhibit digestive enzymes. Studies with trout have shown that the toxin persisted in body tissues 12 months after its feeding to the fish. Much of the retained gossypol was found in the liver, kidneys and spleen. On the other hand, glandless (gossypol-free) cottonseed meal has been used at up to 40 per cent in diets for salmon without deleterious effect.

Binding of gossypol in the oilcake or meal renders the product non toxic. This can be achieved by treating with water and cooking with steam. The gossypol effect may be counteracted by the addition of ferrous sulphate, although the level of additions required has not been established for fish. The range for land species varies from 1 part ferrous sulphate to 1 part free gossypol for pigs to 4 parts to 1 part for poultry layer birds. Only free gossypol which can be extracted with aqueous acetone is physiologically harmful.

Although the amino acid profile of cottonseed protein is somewhat similar to that of soybean and superior to that of groundnut, the danger posed by high gossypol content makes it unwise to use cottonseed meal beyond 15 per cent in fish diets. The rather high fibre content of cottonseed meal also restricts its use.

Cottonseed meal is available in limited quantities in the majority of the countries surveyed, except in Egypt and India, where there are large annual crops of cotton and the feedstuff is the principal protein source for feeds. Cottonseed meal is generally cheaper than soybean meal.

Cotton Seed Oil Cake

Protein content 29 – 42 per cent good source of thiamine and Vitamin E. Presence of polyphenolic pigment gossypol and cyclopropenoic fatty acid adversely affect its nutritional value.

Sunflower (*Helianthus annus*)

Sunflower is a fast-growing cash crop grown primarily for its oil. The leaves and stems of the plant, at two months, are rich in protein and low in fibre and are suitable as feed for fish. There are no known phytotoxins present in the sunflower plant.

Sunflower seed cake made from dehulled seeds has fairly high protein content. It is also richer in the sulphur amino acids than soybean meal although its lysine content is much lower. Sunflower seed meal has been shown to be a good substitute for fish meal in diets for tilapia.

Para Rubber (*Herea brasiliensis*)

The seed of the rubber tree has only been exploited recently as a protein source for animal feeding. The release of prussic (hydrocyanic) acid in raw rubber seeds is prevented by wet-heat processing, which destroys the enzyme that produces it from a glucoside present in the seed. The high temperature associated with expeller cake manufacture also destroys the enzymes.

Decorticated rubber seed oilcake has a composition somewhat similar to coconut oilcake. No data on its amino acid composition is available. Use of rubber seed oilcake in fish feeds has not been documented although it appears that its use will be safe at or below the level used in chicken feeds (between 10-15 per cent of the diet).

Rubber seed oilcake is available in rubber producing countries such as Brazil, Sri Lanka, India (southern states), and the South East Asian countries.

Linseed (*Linum usitatissimum*)

Linseed, or flax, is grown for the production of fibre and linseed oil–a drying oil used widely in paints. The oilcake has a protein content comparable to cottonseed oilcake but is considerably richer in methionine. Linseed cake, like rubber-seed oilcake, contains an enzyme that releases prussic acid from a glucoside present in the seed. Normal processing of the seed for oil removal destroys this enzyme. However, it contains other toxins of unknown identity that cause depressed growth in poultry, even when fed at low levels (3 per cent of the diet). The toxin may have a more pronounced effect on fish since the requirement for pyridoxine, a B-vitamin very much associated with high-protein diets, appears to be greatly increased in rations containing linseed oilcake.

Linseed oilcake is available in considerable quantities in India where it is safely fed to cattle and in lesser quantities in Thailand. Use of the material in compound fish feed in these countries has not however been documented.

Sesame, or Gingelly (*Sesamum indicum*)

After the extraction of oil from sesame seeds, a valuable high-protein feed is obtained. The oilcake is highly acceptable to livestock as well as fish. Experiments on the rohu (*Labeo rohita*) in India have shown that up to 50 per cent of gingelly oilcake may be fed in a complete diet with good results.

Although sesame oilcake usually has higher fibre content than groundnut oilcake, the protein content of both feedstuffs are about the same. Furthermore, sesame oilcake is very rich in methionine (in fact, the richest among oilcakes and meals). Its lysine content, on the other hand, is lower than that of groundnut oilcake. Sesame oilcake and groundnut oilcake used in combination can substitute for the more expensive animal proteins such as meat meal or fish meal. Effects of excessive use of sesame oilcake have not been studied in fish although in conventional livestock, it has been shown to have a laxative effect. Also because of its high phytic acid content, supplemental phosphorus will also be required.

Sesame oilcake is available in fairly large quantities in India, Thailand and Venezuela, and prices are generally lower than soybean meal but higher than cottonseed meal or groundnut oilcake.

Nutrients of Plant Feeds (per cent)

Feed	*Moisture*	*Crude Proteins*	*Crude Fats*	*Crude Fiber*	*Non-nitrogenous Extract*
Soybean cake	11.3	39.1	7.1	4.5	32.0
Peanut cake	11.3	38.4	8.2	5.8	29.5
Cottonseed cake	9.3	35.0	6.0	10.1	30.3
Rapeseed cake	11.0	31.0	6.7	8.2	31.1
Rice bran	11.8	10.8	12	8.2	47.0
Wheat bran	13.1	10.9	3.7	8.9	55.3
Soybean	11.2	38.1	13.1	4.1	27.5
Barley	14.5	10	2	4	69.0
Maize	12	8.5	4.3	1.3	71.0

Food Coefficient of Several Common Feeds for Referential Use

Foods	*Species*	*Food Coefficient*	*Foods*	*Species*	*Food Coefficient*
Snails	Black carp	40	Fresh silkworm pupae	Black carp	3.5
Snails	Common carp	50	Dry silkworm pupae	Black carp	1.5
Clams	Black carp	80	Dry silkworm pupae	Common carp	2
Clams	Black carp	60	Aquatic grass	Grass carp	90
Soybean cake	Black carp	3	Aquatic grass	bream	100
Soybean cake	Common carp	3.5	Land grass	Grass carp	40
Rapeseed cake	Black carp	4.0	Land grass	bream	45
Rapeseed cake	Common carp	4.5	Duck weeds	Grass carp bream	50
Peanut cake	Black carp	3	Rye grass	Grass carp	25
Peanut cake	Common carp	4	Rye grass	bream	30
Cottonseed cake	Grass carp	6	Sudan grass	Grass carp	40
Rye	Black carp	4	Pumpkin vine	Grass carp	35
Rye	Grass carp	3	Wheat bran	Common carp Tilapia	4
Barley	Black carp	4	Rice bran	Common carp Tilapia	3.5
Barley	Grass carp	3	Bean dreg	Grass carp bream	25

Composition of Common Animal Food (per cent)

Composition	*Moisture*	*Crude Protein*	*Crude Fats*	*Non-Nitrogenous Extract*
Kinds				
Fish meal	10	59	9.8	0.4
Fresh silkworm pupae	-	17.1	9.2	-
Dry silkworm pupae	7.3	56.9	24.9	4
Snail	75.8	14.1	0.4	-
Corbicula	85.0	5.3	2	7.0

(H) Animal Based Major Nutritive Ingredients or Feeds of Animal Origin

These ingredients are either from terrestrial, avian or marine animals. They constitute the most important (and often the most expensive) ingredients of aquaculture feeds. Animal protein is necessary to balance the amino acid and vitamin deficiencies in cereals and other plant products. Animal proteins appear to contain unidentified growth factors for some animals. Marine proteins have always been important components of aquaculture diets although shortages of fish meal are stimulating research on methods of replacing them, either partially or completely, with other ingredients.

Ingredients of marine origin are important sources of poly-unsaturated lipids (PUFA's), particularly of the important n-3 series. Some examples of ingredients in this category are blood, feather meal, poultry by-products meal, fish meal, meat meal, raw fish, fish oils, fish silage, shrimp meal and milk by-products.

- ☆ *Meat Waste*: Only available where offal is not valued for human food; may be contaminated and spoiled before use; may also contain animals condemned as unfit for human consumption. Meat scraps and trimmings after fat removed are dried into meat meal. This is inferior to fish meal or soybean meal and is prone to damage (evidenced by low available lysine) by poor processing. If refrigeration is available or the waste can be collected daily both it and blood are potential moist feed ingredients.
- ☆ *Blood*: Blood is often widely available and can be used fresh or as a dried meal. Dried blood meals are often poorly

produced, however and the protein may be damaged. Blood has an extremely high protein content, but does not have such a good amino acid profile as meat waste. The digestibility of its protein is high but its leucine/isoleucine ratio is unbalanced.

- *Bone Meal*: Bone is sometimes mixed with meat meal to form 'meat and bone meal'. Some bone meals contain quite high levels of 'crude' protein but the nitrogen detected in the analytical test comes mostly from indigestible collagen. Bone meal is not usually needed as a calcium and phosphorus source in aquaculture feeds but may become so as less fish and meat meals are used in them.
- *Liver Meal*: A source of B group vitamins, if available.
- *Poultry By-Products*: poultry feet, heads and undeveloped eggs which, like other slaughterhouse wastes are of no value as human food locally, can be used fresh or as a dried meal. Gizzards and intestines can also be used if their contents are removed. Similar in value to meat meal. Very rich in choline.
- *Hydrolysed Poultry Feather Meal*: Feathers hydrolyzed by cooking are a highly digestible source of nitrogen for ruminants. However, as it is severely deficient in most EAAs this product can only be used in aquaculture feeds in small amounts if well balanced by other, good quality, protein. It is believed to be of more value in crustacean than in fish diets.
- *'Trash Fish'*: 'Trash fish' is difficult to define because it differs from location to location in species composition. It consists of fish which are too small, too large, or disliked to an extent to make them unmarketable. What is 'trash' fish in one location may be highly prized elsewhere. What is 'trash' when in super-abundance in one season of the year may be scarce and more valuable at others. Often, it is used for fish meal production and may be mixed with crustacean waste. It is much used as an aquaculture feed alone, or mixed with other materials, or as part of an extruded moist feed. It must be used fresh or refrigerated. Unless pasteurized, some species especially contain the enzyme thiaminase which destroys vitamin B_1, in the other

dietary components. Trash fish is a valuable aquaculture ingredient which is becoming increasingly exploited. Local analysis is necessary before it can be accurately formulated into a compound diet.

- ☆ *Fish Meal*: This is either made by crude processes locally or forms part of the quite sophisticated manufacturing process of an international feed commodity. Fish meals vary widely in their analysis according to the nature of the raw material and the method and care by which it is processed. White fish meal (*e.g.*, South African, Scottish) is made from whole non-oily fish and fish residues which are oven dried and ground. Dark coloured fish meals involve cooking the raw material, pressing it to remove the oil, drying the residue in a steam jacket and grinding it. Shark and dogfish need to be processed by the latter method as they are of little value unless cooked. Peruvian and that fish meals are of the dark variety but differ greatly. That fish meals are made from the whole by-catch (mixed species); Peruvian fish meal is made from one species, the anchovy, and is much higher in its crude protein content. Fish meals are often overcooked, causing damage to the quality of the protein, or undercooked, causing contamination with Salmonella bacteria. Fish meals are often adulterated with urea (which, with a nitrogen content of 46 per cent, has an apparent (N × 6.25) crude protein level of 288 per cent), to make low grade products seem like high protein fish meals. If of good quality, there is no better high protein ingredient readily available for aquaculture feeds. Fish meals should have less than 2 per cent salt and should contain an antioxidant. They are also a considerable source of poly-unsaturated fatty acids, particularly of the higher members of the n-3 series. The key to the successful use of fish meals is knowledge of the processes and quality control standards used by the factories producing them and regular analysis by the user for component quantity and quality. Fish meal is a costly ingredient.
- ☆ *Fish Solubles*: this product, which is the watery material remaining after the oil is removed from the substance pressed out during the manufacture of brown fish meals,

can be condensed and sold separately as a liquid (condensed fish solubles) if not mixed back into the press cake, or dried to form dried fish solubles. Much of the protein of this product is in a non-protein form, but it is sometimes used in small quantities in aquaculture feeds as an attractant. It is high in the B group vitamins and contains an unidentified growth factor for poultry.

- *Fish Silage*: *See* in later pages.
- *Shrimp Meal*: a dried meal similar to fish meal, made from the waste heads and shells of large prawns or shrimps, or from whole small shrimps or crustacean of no human food value. Its true protein value is only about 50-70 per cent (depending on the proportion of heads to shells in the original material) of the apparent or 'crude' protein content. This is because much of it derives from an indigestible (nitrogen containing) polysaccharide, chitin. However, it is an important source of this chitin for shrimp feeds, it is high in choline and it is used for pigmentation as it contains important carotenoids. Both shrimp and fish meals, unless finely ground, give poor stability to aquaculture feeds. Again, as with fish meals, it is important to know the source of the material and to analyse it. Some meals are nearly all shells, with little value. Waste shrimp heads and shells can, if available fresh or refrigerated, be used as excellent ingredients in moist aquaculture feeds, especially for shrimp.
- *Squid Meal*: If available, this is an excellent ingredient for shrimp feeds, but expensive. It appears to have growth promoting properties. Fresh squid can also be used in moist diets.
- *Molluscs*: As squid meal, it can also be utilized fresh in moist diets.
- *Snails*: Can be of value, if cooked and dried.
- *Silkworm Pupae*: Where available in good quantity, this is a valuable ingredient, especially for shrimp feeds. The lipid is prone to rancidity. Only 75 per cent of the total protein is available because, as in the case of shrimp meal, it is

partially chitin. Cocoons have no value.

- ☆ *Milk By-Products*: surplus or damaged milk powder is sometimes available and, if its cost is acceptable, of potential value in aquaculture diets. Whey, a residue for the production of cheese, as well as a high-protein waste from the refining of animal ghee, are utilizable up to a 10 per cent inclusion level, at least for salmonids. Whole milk powder has an EAA profile which is generally regarded as better than fish meal and close to the 'ideal' food, chicken egg protein. Skimmed milk has a similar EAA profile.

Meat Meals and Raw Slaughterhouse Waste

Animal protein supplements, or meat meals, are the principal by-products of animal slaughterhouses. They include meat scraps and trimmings after the fat has been removed during the rendering process, as well as internal organs of slaughtered animals. Their availability as feed depends in part on the food habits of the human population.

The quality of meat meal as a protein supplement depends on its production process as well as raw material used. Good quality meat meal is low in rat and mineral content. Meat meal is frequently used as an animal protein source in compound fish feed manufacture although its feed value is generally considered inferior to that of soybean meal and fish meal. Fats in meat meal although poor in the PUFAs required by cold water species, are good energy sources for both carp and rainbow trout.

However, because of their high water content, slaughterhouse wastes are best used in combination with dry feed ingredients to make moist feeds. Heat processing may be necessary and the material used immediately if refrigeration facilities are not available. Because meat meal is usually an imported item, its cost is usually high. Slaughterhouse waste, on the other hand, is inexpensive when available.

Blood

Animal blood is obtainable in most areas from abattoirs. In many village abattoirs, it is merely discarded. Animal blood may be used fresh, or processed into blood meal before using. Although dried blood has a high-protein content and is highly digestible, its amino acid content is not as well balanced as that of muscle tissue. Fresh blood may be used to enrich commonly available feedstuffs such as rice bran and wheat bran in fish diets. The bran also soaks up the excess moisture in the production of moist pelleted feed. In combination with rumen contents, fresh bovine blood has successfully substituted for fish meal in channel catfish diets. Because animal blood is widely available in most countries, often at little or no cost, its use in aquaculture diets can help lower feed cost. Blood meal is superior to meat meal in test-diets for salmon.

Bone Meals

Bone meals are made by heat processing of animal bones followed by grinding to produce a fine powder. Contrary to common belief, bone meal is not all calcium and phosphorus. Depending on the processing method employed, bone meal can contain up to 36 per cent protein. Much of this protein is, however, of low quality since it is mainly collagen, the substance extracted industrially from bones to produce gelatin and glue.

With the development of fish diets containing ever-decreasing content of animal protein such as fish meal and meat meal, bone meal may become more prominent as a dietary source of calcium and phosphorus.

Hydrolyzed Poultry Feather Meal

Prepared from poultry viscera, feet, heads and undeveloped eggs. Contain 45 – 60 per cent protein. It is a better source for prawn than fish diets.

Trash Fish

By-catches of the fishing industry have traditionally been used as raw materials for fish meal production or as feed for livestock and fish. The composition of such landings varies with the geographic area where the fish are caught. In tropical waters in the Indo-Pacific region, the threadfin snapper (*Lutianus nematophorus*) Malaysia, and the silverbelly (*Leiognathus splendens*) India, Indonesia, Sri Lanka, are among the principal species caught. Freshly landed trash fish have a high feed value.

Trash fish, when fed fresh, with or without prior cooking, are superior to fish meal in terms of protein content (expressed as percentage dry matter) and protein quality. When mixed and properly processed in correct proportions with less expensive, but more readily available ingredients, such as rice bran and with adequate supplementation of vitamins, trash fish make excellent diets for aquaculture. A major problem associated with feeding trash fish is maintenance of water quality. This can, however, be partially overcome by proper choice of the dry components, *e.g.* rice bran which absorb the excess moisture and appropriate diet preparation methods to produce feeds that remain stable in water until consumed.

Unlike most other ingredients, supply of trash fish is highly seasonal. When available, trash fish are considerably less expensive than fish meal.

Fish Meal (FM)

Fish meal is perhaps the most abundant animal protein source commercially produced and marketed in several countries. The quality of FM depends on the species, size, maturity age and process employed in manufacturing. FM is a by-product of the fisheries industry in which whole or cuttings of fish are cooled and dried. The methods of preparation of FM include–Vacuum dried and Sun dried – best one. Flame dried – not recommended as it render protein less available, oxidizes lipid and produces antinutritional factors *i.e.* histamine.

Best method – Stem cooking. Pressing to remove water and oil drying.

FM is highly palatable to shrimp and serve as desirable attractants because it contains Inosine, Inosine PO_4, certain L – amino acids. Brown fish meal–Clean, dried ground tissue of undecomposed whole fish or fish cuttings, of both, with or without extraction of oil.

White fish meal–dried, non-rendered, clean, undecomposed protein of fish from fish processing waste typically from whole fish. White fish meal is better than Brown fish meal based on quantity and quality of lipids in respective meals. The protein content in fish meal ranges from 50 – 65 per cent Fish meal should contain a minimum of 60 per cent Protein. Omnivores – 5 – 15 per cent; Carnivores fish feed – 20 – 40 per cent In commercial prawn feed, FM levels range from 10 – 40 per cent The quality standards for fish meal are:

Crude protein	> 68 per cent
Lipid	< 10 per cent
Ash	<13 per cent
Salt	< 3 per cent
Moisture	< 10 per cent
NH_3-N	< 0.2 per cent
Anti-oxidant	200 ppm
Size	finer than 0.25 mm
Processing	Steam processing

Fish meal remains an important, but expensive ingredient in most fish and crustacean diets. It is especially rich in the essential amino acids such as lysine and methionine and minerals and is highly digestible for fish. Its use appears limited only by high cost and availability. Dry diets containing a high proportion of fish meal are difficult to pelletize.

Fish Solubles (FS)

Fish solubles are the water material remaining after the oil is removed from the liquid pressed out during manufacture of fish meal. High in B group vitamins and UGF. Serve as attractants.

Fish Oil (FO)

Fish oil is a by-product of fish meal production. It is rich in vitamins A and D and polyunsaturated fatty acids that are required by fish. Although widely used to supply these nutrients in fish diets in Western countries, fish oil are seldom used in commercial fish feed manufacture in developing countries. The cost of fish oil depends in part on food habits of the human population. In some countries, the oil from fish meal production is merely discarded.

Fish Oil

PV	< 6 mg/kg
Anisidine Value	< long/lr.
Total pesticides	< 0.4 ppm.
PCBs	< 0.6 ppm.
N_2	> 1 per cent
Moisture	< 1 per cent
Antioxidant	500 ppm

Fish Silage

Fish silage has not come into widespread use as a feed for fish. Although developed upon the need for a less expensive method for preserving trash fish, it is also a cheaper alternative to fish meal for use in aquaculture diets. It is used in commercial salmon farming in Denmark and Norway and in experiments in developing countries involving warm-water species, *viz.*, the snake-head in Thailand and the carp in Indonesia.

Fish silage can be prepared in a variety of ways. These include fish/carbohydrate fermentation and ensiling with mineral and/or organic acids.

Acid silage in prepared from ground trash fish waste fish head, viscera prawn waste small crabs and mixed with a mixture of acids such as HCl/H_2SO_4 and propionic acid or ferric to cause liquefaction and to prevent bacterial decomposition. Biologically, fishy silage in preferred by introducing lactic acid bacterial into a ground fish CHO mix. The visually liquid product can be used as ingredients mostly in fish forms.

Crustacean Meals

Meals obtained fray small prawns, prawn heads, mantis shrimp, crab, krill. Good attractants for prawn. Incl. Level: 5 to 15 per cent

Shrimp Meal

Shrimp meal is made from freezing-plant wastes (heads and shells) or from whole shrimps not suitable for human consumption. The material is first dried in the sun or in an oven and then ground. Because shrimp exoskeleton is primarily chitin, an indigestible nitrogen-containing homopolysaccharide, the true protein content of shrimp meal made from freezing-plant wastes is only 50 per cent of the value obtained by proximate analysis. This high content of indigestible matter restricts its use to less than 25 per cent in aquaculture diets, Shrimp meal is especially rich in choline, a B-vitamin. Because crustacean also contain carotenoid pigments, shrimp meal has long been used in trout and salmon diets to improve flesh colouration in the fish. Up to 15 per cent in the diet has been used for this purpose. The exoskeleton of shrimp is tough. Unless finely ground, shrimp meal added in large amounts to the diet may result in difficulty in obtaining good quality pellets. Shrimp meal, where available, is usually far less expensive than fish meal.

Molluscan Meal

Sources: Clam, mussel, snail, squids, cuttle fish. Incl. Level: 5 to 10 per cent

Silkworm Pupae and Other Insect Larvae

Available in good quantity only in India and Thailand; silkworm pupae are rich in nutrients. When fed fresh they are a rich source of protein as well as fats (especially unsaturated fatty acids). The meal, from which the fat has been extracted, has been used with better results as a fish meal substitute in experimental diets for carp. Due to the high content of chitin in silkworm pupae, its actual protein value is only 75 per cent of the value obtained by proximate analysis.

It can be used as an ingredient in fish and prawn feeds at low level. Higher level in phone to rancidity for this, solvent extraction of lipids may improve the quality. 72 per cent crude protein an dry weight basis. It is used to feed carnivorous fishes. Oil can also obtain from these pupae called chrysalis oil (which is range as fish oil). This oil is used as lipid source in aqua feeds.

In Japan and China, silkworm pupae are usually applied to feed fish. Fresh pupae are more effective, but difficult to preserve, whereas dry pupae contain rich fats, but these fats are easy to deteriorate by oxidization. Furthermore, the taste of fish flesh is not so good; therefore, pupae feeding must be stopped 2 to 3 weeks in advance before harvest.

Milk By-products

Milk by-products such as milk powder, due to their high cost, are seldom used in preparing fish feed, although the milk protein, casein, is widely used in purified diets for experimental purposes.

Whey and dairy-processing wastes have been successfully used at the 10 per cent level in diets for rainbow trout. Their incorporation into diets also improves pellet quality. Whole milk powder or skimmed milk powder has amino acid profile close to that of chicken egg protein. It can be used in feeds if available at reasonable cost.

Chicken Egg

Without shell, it has about 46 per cent Crude protein; 43 per cent Lipid and 4 per cent ash. It is used as an ingredient in hatchery and nursery feeds for fish and shell fishes.

(I) Miscellaneous Feed Stuffs (Unconventional or non-conventional feed ingredients)

Many other ingredients have potential in aquaculture feeds; their value has not yet been fully evaluated. Some of these ingredients are referred to as 'unconventional' or 'non conventional' feedstuffs.

This group of feedstuffs includes leaf protein concentrate, minerals, seaweed, by-products of sugar cane, by-products of fermentation processes, lipids, microbial proteins, algae, manures and celluloses.

Seaweeds

Seaweed meals are of value in aquaculture diets, particularly in those for shrimp. They are a source of trace minerals and vitamin A and probably increase palatability. Shrimp are able to digest cellulose better than fish.

Manure

Animal manures and litter wastes can be used in fish and shrimp feeds in addition to their use as manures. However, over 50 per cent of the crude protein of manures consists of non-protein nitrogen, such as uric acid. It is believed that in most cases where manures have been used as an ingredient in compound aquaculture feeds the true effect has been to fertilize the natural food in the pond. However, use of small quantities of poultry manure in shrimp feeds was found successful and tilapia was experimentally fed with a trout feed having up to 30 per cent dried poultry waste in it, without any depression of growth rate.

Non-Protein Nitrogen (NPN)

Although there are conflicting reports about the ability of some species of fish and crustaceans to utilize the sources of NPN, such as urea, biuret and ammonium phosphates, it has not yet been recommended that these materials can be used in fish feeds.

Microbial Proteins

Bacterial proteins, grown from aqueous solutions of mineral salts with a nitrogen source, using methane as an energy source, and yeasts (mostly 'torula' or fodder yeasts) grown on paraffins and industrial wastes, such as molasses, sulphite waste liquor from the paper industry, fruit wastes, etc., are becoming increasingly available as ingredients for animal feeds. Both types, especially the bacterial type, are rich in protein. The presence of large quantities of nucleic acids limits their use as human food, but it is unknown if this is a problem in their use for fish feeds. The amino acid profile varies according to the type and the media on which it is grown. Some yeast is particularly deficient in methionine, while some are very high in lysine level.

Bacterial Protein

Bacterial protein is among the single-celled proteins (SCP) that have received increasing attention in recent years. *Methanomonas* spp. is cultured in an aqueous suspension containing mineral salts and a nitrogen source. A mixture of methane and air is bubbled continuously through the suspension. The harvest of bacteria, usually one in every three days, has a high protein content, a good portion of which is nucleo-protein. Although its production may be feasible in countries which have a surplus of methane, the cost of the product is presently high. Studies on its use in fish feed are not conclusive although there are indications that it is superior to other SCP such as yeast and algae.

Algae

Dried unicellular algae–like microbial protein, sometimes referred to as single cell protein (SCP) (mostly *Chlorella, Spirulina* and *Scenedesmus* species)–are high in protein but usually too expensive for use in animal feeds. Those fed alive to larval aquaculture stock are especially cultured. Algae are rich in carotenoids.

The cultivation of micro-algae as food or feed is not new. The production of algal blooms and benthic algal complexes through fertilization and water management has been long practiced in Asia. Certain species of *Chlorella, Scenedesmus* and *Spirulina* have been established as excellent feed for larvae of many cultured species of fish. Large-scale mono culture of these SCPs, however, involves technology that is still at a developmental stage. Moderate scale production at the rural level employing animal (particularly pig) manure has been shown to be feasible. The harvested product, when dried, is non-toxic and can be used to supply practically all the protein required without ill effect, although supplementing lysine (for *Chlorella*) or methionine (for *Scenedesmus* and *Spirulina*) may be necessary. *Scenedesmus obliquus* has been tested quite extensively in carp and found to be a good substitute for fish meal.

Yeast

The most commonly used and widely available SCP for animal feeding is torula yeast (*Candida utilis*). This is cultured on substrates comprising a variety of industrial wastes, including: molasses, dried citrus pulp, or sulphite liquor from the wood pulp and paper industries. The harvested yeast is usually dried over steam-heated drum rollers. More recently, paraffin-grown yeasts have been produced in small quantities for evaluation as food for both livestock and humans. Feedstock for production of such yeasts is diesel oil or n-alkanes.

Feed-grade yeast has been shown to be excellent substitutes for fish meal at low levels in diets for both fish and livestock. In general, yeasts are relatively low in methionine. Proper supplementation with synthetic sources of the amino acid could permit yeast to be the only protein source in the diet. On the whole, petroleum yeasts (*Candida lipolytica*) have higher feeding value than yeasts made from molasses (*C. utilis*). Yeasts, like bacterial proteins, are rich in nucleic acids which limit its use as food for humans. Whether this applies also to fish has not yet been determined.

Leaf Protein Concentrates (LPC)

Leaf protein concentrate is produced by first grinding the plant leaves and then separating the juice by pressing. The juice, which contains the dissolved protein, is then coagulated by heating. The curd obtained is removed and dried.

The machinery required for large-scale production of LPC is expensive; the minimum economical output being 10 tons of leaf protein per hour. Smaller units are being designed for use at the village level.

Dried LPC has protein content somewhat lower than that of soybean meal. Its protein quality, on the other hand, is higher than that of the latter, but lower than that of fish meal. LPC made from the genera *Leucaena* and *Mimosa* contains toxic mimosine, a cyclic amino acid. Although few studies have been conducted with fish, it is presumed that LPC made from alfalfa and other legumes may be safely used. With other sources, the presence of toxins other than mimosine, viz., haemaglutinins, glucosides, saponins, etc., have to be taken into account. Leaf protein concentrates can be produced from the leaves of tropical legumes such as the *Acacia* and *Desmodium* species and from grasses. Leaf protein concentrate from rye grass used at levels of up to 48 per cent of total dietary protein have been successfully tested on carp and trout.

Lipids

Many different natural animal tallows (solid above 40° C), lards (melting point between 20-40° C) and oils (liquid below 20° C) are available. Generally, tallows come from cattle or sheep, lard from pigs, horses and any type of bones and oils from marine animals. The vegetable lipids which are available are also mainly oils. Processed animal lipids are not always available but vegetable oils are; however they may be very expensive because they are refined for human food. They may thus be quite an expensive source of energy and EFAs for livestock. Soap stocks, materials which are used in the production of soap, consist of the sodium salts of free fatty acids and traces of protein. Some also contain carotene pigments. Soap stocks can be used as sources of energy and fatty acids in feeds.

Fish oils and most vegetable oils are high in PUFAs and need to have antioxidants added to them during processing to delay the onset of rancidity. Hydrogenated (hardened) fats made from fish oils and beef tallow are poorly digested if their melting point is above 40°C. The main difference is that the marine oils have significant levels of HUFAs (n-3 series) which are absent in vegetable oils.

Most common are oils from coconut, soybeans, groundnut and palm nut. Even then, these are usually refined for human consumption and with the exception of crude palm oil are too expensive to use in feeds. Whenever available, however, their inclusion enhances the feed value of fish diets significantly.

The following tables gives the PUFA content of fats from a wide variety of animal and plant sources.

Fat	*Per cent PUFA*
Animal or fish sources	
Lard	11.8
Beef tallow	4.2
Cod, Atlantic	42.8
Herring, Atlantic	14.6
Rainbow trout	31.0
Common carp	22.5
Shrimp	41.6
Vegetable sources	
Rice	50.0
Maize	58.2
Wheat	60.5
Groundnut	31.0
Sesame	40.5
Cottonseed	50.7
Soybean	57.6
Sunflower	63.8
Safflower	73.8
Palm	9.3
Olive	9.0
Coconut	2.0

Cane Molasses

This diluted by-product of cane sugar refining can be used as a partial source of energy, as "filler", as a binder and as a possible attractant. It is cheap and widely available, except where fermented for alcohol production.

Refining of cane sugar produces a by-product known as blackstrap molasses. This highly viscous (specific gravity 80-85°C Brix) material is difficult to handle at temperatures below 30°C. Therefore, blackstrap molasses is normally diluted with water to lower the specific gravity to about 79.5° Brix, so that it may be more easily transferred by pumping.

Cane molasses predominantly sugar of which roughly half is sucrose and the other half is of reducing sugars. Its principal use in the production of compound feed for poultry and livestock is as a dietary energy source, although it also enhances acceptance of the feed. Its incorporation into pelleted feed has a positive effect on the physical properties of the pellets.

Use of molasses in compound fish feed has not been thoroughly investigated. Presumably, its level of usage should not exceed 5 per cent of the dry diet, considering its established laxative effect (when used beyond 5 per cent for poultry and 15 per cent for pigs). Because of its relative cheapness and wide availability in the tropics, molasses is a good partial replacement for the more expensive cereal grains as an energy source. Because molasses ferments rapidly, it may also be added in small quantities along with rice bran to trash fish in the production of fish silage.

Breweries

1. Brewer's grain waste (extraction of malt for production of beer).

   Dried brewers grain contains 19 per cent protein; 3.5 per cent fat tried in Freshwater prawn and Nile tilapia.
2. Brewers yeast (40 – 45 per cent protein).

Brewers Spent Grains

This material, the extracted malt, is a valuable source of protein and energy which is often thrown away or used as a fertilizer. Brewers spent grains, or brewery wastes, as they are often termed, are residues from the initial breakdown of starch to fermentable sugars in beer manufacture.

Fresh brewers grains are fairly rich in protein and digestible carbohydrates. The material, if mixed with other feedstuffs before feeding, should first be boiled to deactivate enzymes that may otherwise act on carbohydrates to produce alcohol, and then organic acids.

Raw brewery waste is available in considerable quantities in large urban areas of most countries. Levels of up to 50 per cent in the diet have been successfully fed to tilapia in Africa.

Brewers Yeast

Brewers yeast is obtained as a residue after the fermentation process in beer production. Yeast is a rich source of B-vitamins (excepting vitamin B_{12}) but low in calcium. Due to its high cost, it is seldom used at levels beyond 5 per cent of the total diet.

Dried brewers yeast, if available, is an excellent raw material. It is a by-product of beer production. Like brewers spent grains, the fresh material requires boiling to inactivate the enzymes present before use as a wet ingredient. It is a rich but expensive source of B group vitamins and protein. This yeast is usually *Saccharomyces* spp.

Grain Distillers By-products

These by-products of ethanol or acetone butanol production vary in analysis according to the method of production and the type of raw cereal used. They are generally of greater value than brewery wastes and more expensive. They appear to stimulate growth in poultry but it is not known whether they would be as effective in aquaculture feeds. This is useful if economically available. Distillery grain waste offers potential.

26 – 30 per cent crude protein; 9.2 per cent fat; 9 per cent crude fibre. Trout diet (up to 21 per cent); channel cat fish (up to 40 per cent). Deficient in lysine; but supplemented with crystalline *L. lysine.*

Distillery wastes are obtained from the manufacture of alcoholic spirits. Their composition depends upon the type of grain used, *viz.*, maize, sorghum, wheat or rice. Two processes are employed in making alcoholic spirits. The first is known as the British process which involves the two basic steps of grain conversion, followed by separation of spent grains, fermentation of sugars to alcohol and final distillation of the product. This process yields two products: distillers grains, and distillers solubles (including yeast). The other process, known as the American process, does not involve separation of spent grains before the fermentation process and yields only one product that includes both the spent grains and yeast.

Distillers by-products have lower fibre content than brewery waste. The material, when added to rations, appears to stimulate growth in poultry and livestock. The effect has been attributed to unidentified growth factors believed to be present in the material. Due to its lower availability and higher feed value, distillery wastes are usually more expensive than brewery wastes. The material is used whenever available at low levels in compound fish feed.

Brewer yeast is the dried sterilized, unextracted yeast (*Saccaromyces*) resulting as a by product from the brewing of beer.

Crude Protein: 47 – 45; Crude Fibre: 2 – 7 per cent; Ash: 6 – 9 per cent.

Lactic yeast is a good ingredient for prawn feed. (2 to 5 per cent level inclusion).

Starch Industries

Maize gluten and wheat gluten use the by products from grains to wet milling operation gluten (45 – 48 per cent protein); Tilapia (16 – 49 per cent); wheat gluten (12 per cent level as binder).

Sugar Industries

1. Molasses (Less than 3 per cent Protein): 1 – 9 per cent level in fish diet due to stickiness. At higher level 25 per cent and above, has laxative effect.
2. Press mud – by product of sugar cane industry during precipitation 25 per cent level replacing Rice bran in traditional diet. 15 per cent protein and oil (7 per cent) apart from minerals.

Oil Industries: Cakes and Meal

Conventional	*Non-Conventional*
– Ingredients on costlier	– Not costlier
– Contain more per cent of protein	– Lesser per cent of protein
– Cost of production/kg of feed is high	– Cost of production/kg of feed is less

Self Assessment Questions

1. What are feed ingredients?
2. Compare Nutritive feed ingredient and Non nutritive feed ingredient (additives).
3. Compare energy and non energy feeds.
4. List out the different feed ingredients available in your area.
5. Classify generally the feed ingredients.

Chapter 13
Feed Additives

It is also called as non-nutritive feed ingredients. A variety of substances are added in aquaculture feeds to protect the labile nutrients, to improve nutrient availability and utilization by the animals. When additives are added in small amounts, it improves tremendously the performance and efficiency of the feed by 10 – 25 per cent. However, though they improve growth rate, survival and FCR, their cost effectiveness should be evaluated before they are incorporated in commercial feeds. The following are the additives used in aquaculture feeds:

1. Pellet binders
2. Antioxidants
3. Preservatives
4. Chemo-attractants and feeding stimulants
5. Pigments
6. Anabolic agents
7. Probiotics
8. Immunostimulants
9. Micronutrients
10. Miscellaneous additives

1. Pellet Binders

It provides desired water stability to the feeds. It also prevents disintegration of feeds and leaching of nutrients into the water. Water stability of feeds can also be improved by the use of very finely ground raw materials and thicker small hole die plates.

Pellet binders are divided into nutritive binders (Tapioca, pregelantised starches, wheat gluten, cottonseed meal) and non-nutritive binders (CMC, alginates, agar and various gums).

Nutritive binders – pregelantised starches, wheat gluten, cottonseed meal; Non-nutritive binders – Tapioca, CMC, alginates, agar and various gums.

Substances used for inducing water stability in aquaculture feeds are Casein, Gelatin, Collagen, Chitosan, Guar Gum, Locust Bean Gum, GFS (xanthan gum, locust bean gum, guar gum mixture), Agar, Carrageenin, Seaweed binder, Corn Starch, Tapioca Starch, Potato Starch, Wheat Gluten, High Gluten Wheat Flour, Carboxymethylcellulose (CMC), Sodium Alginate plus Sodium Hexametaphosphate, Lignosulphonates, Hemicelluloses, Bentonites, Polymethlolcarbamide (Basfin), Carbopol, Hydrolysed Polyvinyl Alcohol, XB-23 (an anionic heteropolysaccharide).

Ingredients Antagonistic to dry pellet water stability are Lucerne Meal, Cereal hulls, particularly rice and oats, Bone, Salt, Molasses, High lipid levels (> 10 per cent), Brewers grain (high levels), Whey (high levels), Salt and Molasses have been used to increase stability in moist feeds.

The main function of binders is to improve their pelletability, to enhance their durability, to preserve their physical form during storage and to enhance the water stability.

2. Antioxidants

Antioxidants are added in the diet to prevent or delay the onset of lipid rancidity and prevent vitamin loss in the feed. Antioxidants may be natural (Vitamin E–*a* tocopherol and C or synthetic chemicals (Butylated Hydroxyl Toluene–0.2 per cent, Butylated Hydroxyl Anisole – 0.2 per cent and Ethoxyquin–0.015 per cent).

3. Preservatives (or) Antimicrobial Agents

Being highly nutritive under favourable condition, susceptible for the growth of microorganisms (bacteria, yeast and fungi).

Aspergillus spp., *Fusarium* spp. and' *Penicilliun* spp. are associated with this spoilage. They alter the nutritional status, cause bad flavour and taste, staleness and produce highly toxic and carcinogenic mycotoxins (aflatoxins, trichothecanes, ochratoxin and zeralenone, T-2 toxins etc.). Preservatives are useful in controlling fungi. The most widely used preservatives are calcium and sodium propionates at 0.1 – 0.25 per cent. Others used are sorbic acid, potassium sorbate, sodium sorbate, propionic acid, calcium sorbate and calcium benzoate.

4. Chemo-attractants and Feeding Stimulants

It induces feeding behaviour and help to improve feed intake. It helps the fish to recognize, ingest and grasp the feed. Important chemo-attractants are blanch water, fish solubes, glycine betaine, Inosine, Inosine Mono Phosphate (IMP), Free Amino Acids and Nucleotides. Mixtures of L – amino acid, glycine – betaine and Inosine or IMP are considered as universal feeding stimulants for fish.

Squid, shrimp, clam, mussel and polychaete extracts are known to be excellent natural attractants and feeding stimulants for prawns, shrimps and carnivorous fish (sea bass and red sea bream). Feed stimulants possess some general characters such as (*a*) contain N_2, (*b*) low molecular weight (<1000), (*c*) non volatile and water-soluble and (*d*) possess acid and base proteins simultaneously. Free amino acids and nucleotides are most important and these are naturally occurring at very high levels in shrimp meal, squid meal, fish meal, crab meal and clam meal. For shrimp, the important are betaine, glycine and alanine.

5. Pigments (Carotenoid Supplements)

Xanthophylls and carotenoids are the most important classes of pigments for fish and crustaceans. Crustaceans and polychates are very good source of carotenoids. Astaxanthin is added at 50 ppm fed to shrimp for 6 weeks improve coloration and also alleviate the effects of blue shrimp. Fish and crustaceans cannot synthesize the pigments, but can alter the molecules by oxidation. In the case of crustaceans, carotenoids have an important function in reproduction. Addition of carotenoids to broodstock diets resulted in shortening of the maturation period, enhanced hatchability of eggs and better survival of larvae. Addition of carotenoids has many possible functions in fish. They act as an antioxidant, which is 50 times faster than vitamin E and precursor in Vitamin A synthesis.

6. Anabolic Agents

Anabolic agents are added to improve the metabolic or digestive efficiency and to promote protein deposition.

Hormones

Important hormones are 17 methyl-testosterone, thyroxin, insulin, triodothyronine, growth hormone and recombinant bovine somatotropin and thyroprotein.

Feed Antibiotics

Antibiotics are added to feed to treat disease. Routine uses in feed are not recommended. Terramycin (6000 – 10,000 IU/kg), Flavophoslipol, Virginlamycin, Zinc Bacitracin and Inophores are important antibiotics used in aquaculture.

Zeolite

It is nothing but built in toxic metabolite remover. Up to 2 per cent will be included in the diet. It is a naturally occurring hydrated sodium alumino-silicate. It has the ability to bind and remove NH_3. It is also serving as a source of trace minerals.

Enzyme

Proteolytic and amylolytic enzymes are added to feeds to improve protein and CHO digestion respectively. Proteolytic enzymes are important and they are bromelin and papain (0.1 – 0.2 per cent). However, enzyme supplementation to feeds is expensive.

Phytosterols

Phytosterols are steroids, which are extracted from plant products. It contains a variety of steroids including ergo-sterol, stigma-sterol, beta – sito-sterol and diosgenin, but do not contain cholesterol, an essential nutrient for shrimp.

Olaquindox (Growth Promoter)

It appears to partition energy in animals for protein synthesis. Inclusion level is 20g/MT of feed.

Bile Acids

It assist with the assimilation and absorption of lipids and lipid-soluble substances, including cholesterol, phospholipids, fatty acids and fat-soluble vitamins. It is also helpful to maintain proper functioning of the hepatopancreas.

7. Probiotics

Probiotics are additives and alternative to antibiotics. Probiotics means, "favour for life" (pro means favour or promote; bio means life). This term was first used to describe substances secreted by microorganism, which stimulate the growth of other microorganisms. Parker (1914) defined the term probiotics as "organisms and substances, which contribute to intestinal microbial balance". In aquaculture, the action of probiotics is as follows:

1. They are found useful in keeping the balance of the gut micro flora, which benefits the animal, by protection against disease and by improving nutrition.
2. They utilize nitrogen-bearing wastes in water and in the metabolism process, convert them into inorganic nutrients that can be easily utilized by the system.
3. They replace or destroy harmful bacterial population by competitive inhibition. In addition, some of the antibacterial substances produced by certain bacteria, which has been selected, actively kill the harmful organisms mainly belonging to the *Vibrio* family including fungi.
4. High enzymatic activities such as amylase, protease, lipases, etc. remove organic load caused by excess feed, faecal matter and dead algae. The microbes create a perfect environment and balance the carbon and nitrogen cycle.
5. In case of the soil also, the beneficial bacteria helps in downgrading the harmful effects of the formation of sulphides and nitrites due to the heavy organic load accumulated during the culture period.

Both indigenous and imported probiotics are now available in the market. Three different probiotics, which will be highly useful for the aquaculture farmer community for reaping better profits. They are 1. a gut probiotic 2. a water probiotic, and 3. a soil probiotic. Either individually or in combinations, they will form the ideal weapons for a farmer to combat the microbial studies and improve the pond economy. Common probiotics contain *Lactobacillus*,

(*Lactobacillus* species like *L. acidophilus, L. factis, L. plantarum*), *Enterococcos faccacis* and yeast. The most commonly and extensive used one is *lactobacillus*. Probiotics products are available in the form of oral pastes, capsules, water dispersible powder, liquids and feed additives, which include microbial cell, microbial culture and microbial metabolites. The microbial strains used as probiotics, should have the following characters: Non-pathogenic, Non-toxic, Present as viable cells, survival in the gut environment (resistance to low pH, organic acid etc.) and stable and viable for long periods under storage.

8. Immunostimulants

It is an agent (chemical, drug and stressor), which stimulates the non-specific immune mechanism when given alone or the specific mechanism when given with an antigen, and these are natural products. Hence, there are no environmental hazards and they have no residual effect on fish which can be given orally or through feed.

Glucans

It is used either direct injection or food. An experiment conducted showed an increased resistance to *Vibrio anguillarum* and *V. salmonicida* in Atlantic salmon. Laminaran (glucan from *Laminaria*) is also potential feed additive.

Yeasts

Their suitable size and high stability in the water column mean that yeasts can easily be ingested by filter feeding organisms. Yeast can be mass-produced at a relatively low cost. Yeast is used as feed ingredient and contains vitamin B, protein and fatty acids. It is known as an anti-stress factor.

Lectins

Lectins are proteins, which can bind to CHO with or without catalytic activity. They are known to bring about agglutination of foreign particles containing CHO in their membrane.

9. Micronutrients

The iron bearing proteins like transferrin and lactiferrin are involved in immunological responses in the process of a non-specific defense mechanism. In iron deficient animals, antibody production and the activity of natural killer (NK) cells are severely impaired. Other minerals like Cu, Zn and Mn are also involved in multi level defence systems in protecting the cells against chain reactions initiated by free radicals comprising superoxides dismutase.

Miscellaneous Additives

Aspirin is used in tilapia fry at 1g/kg feed level as an anti-stress factor before transferring the fry to seawater raceways. Similarly, sorbitol for liver health, glycerol-oleate for water oil emulsifier, carnitine for better utilization of lipid and sodium polyphosphate used as anti-viral effect.

Feed additives and immunmodulatory feed based additives are one of the tools that farmers and feed manufactures can use safely and effectively to fight diseases. They should be used in conjunction with healthy rearing conditions and adequate nutrition. The addition of probiotics and micronutrients has shown promising results in both experimental and culture conditions. Further, studies would prove the role of immunostimulants in the culture of healthy and disease-free fish.

Self Assessment Questions

1. Are additives used as feed ingredients? If yes, Why?
2. List out the different feed additives used in fish diet.
3. Difference between nutritive and non-nutritive binders.
4. Why binders are added in fish diet?
5. Why anti-oxidants are added in fish diet?
6. Why preservatives are added in fish diet?
7. What is immunostimulants?
8. Why pigments are added in fish diet?
9. What is the importance of using immunostimulants and miconutrients in fish diet?
10. What is gut probiotic?
11. Explain the term eubiosis and dysbiosis.
12. List out the miscellaneous additives used in fish diets.

Chapter 14

Physical Evaluation of Feeds and Feed Ingredients

Introduction

Quality of aqua feed and feed ingredients are tested by evaluation methods. Both the aqua feed and feed ingredients can be tested by the following evaluations:

1. Physical evaluation
2. Chemical evaluation
3 Biological evaluation and
4. Microbiological evaluation

Evaluation of a feeds and feed ingredients are more important than formulation. Evaluation may be done by adopting various procedures.

Uses of Evaluation in General

1. To check on the accuracy of manufacturing process in arriving at a finished feed of desired concentration.
2. To measure the nutrient loss during manufacture and storage.
3. To predict nutritional value of particular formulation.

Physical Evaluation

Advantages of Physical Evaluation

1. No cost is required
2. No skill is required
3. Easy to perform when compared to other evaluation methods
4. Quicker method
5. Minimize adulteration of final product

Disadvantages of Physical Evaluation

It indicates only qualitative aspects and not quantitative one.

The first step in assessing the quality of ingredients and processed feeds are to examine or inspect its physical characteristics. The criteria included in the physical characteristics are:

1. Purity of ingredients–with the aid of microscope one can confirm purity of ingredients (Identifying foreign contaminants or adulterants).
2. Particle size and distribution
3. Density
4. Water stability
5. Texture
6. Feed shape and Pellet quality
7. Homogenity of ingredients
8. Colour and contrast
9. Odour

Among the nine factors listed, the factors such as pellet stability, shape and density is very important because the property greatly influence the extent of feed utilization. The following are the few quicker methods, which help to assess the quality of feed and feed ingredients before purchase.

Appearance

- ☆ Feed should have uniform color and shape.
- ☆ Avoid spots in the feed which indicate improper mixing of ingredients.

- ☆ Avoid dark colours in feeds (which indicates poor quality ingredients or over cooking).
- ☆ Feed should not have any impurities.
- ☆ Feed should be free from fungi etc.
- ☆ There should not be any caking of feed.

Feel

Push your hand into the feed bag and draw it out. Presence of powder sticking to your hand is not good which indicates poor processing of feed. While broadcasting in to fish ponds, the powder would be blown away, reducing the quantity of actual feed and resulting in wastage and inefficiency.

Smell

Fresh fish smell indicates good feed. Smell of heavy oil indicates poor oil quality or poor soya oil used without deodourising.

Taste

Good feed should be slightly salty and give a sweet after tastes, indicating good fish meal. Bad and bitter taste indicates rancidity of oil used. A clean bite (pellet) also indicates that moisture content is low.

Water Stability

Test by putting feed into glass of water and the shape should be maintained at least for two hours. Some colour should have diffused into water within 30 min. It also indicates attractant used in the feed has dispersed.

Determination of Water Stability of Aqua Feeds

It is carried out by two simple methods.

1. Qualitative evaluation
2. Quantitative evaluation

Procedure for Qualitative Evaluation

Take 10 g of pellets in 500 (or) 1000 ml beaker

↓

Add 850 ml of water

↓

Gradually stir with a magnetic stirrer or hand stirring for every 10 min for a fixed time

↓

Look for visual observation (time taken for the feed to disintegrate in water

↓

Reporting of the results as water stability in terms of duration (in hrs)

Procedures for Quantitative Evaluation

Take measured quantity of feed suspended on a 2-3 mm mesh screen repeatedly immersed or submerged in and out of water

↓

After particular time, filter the feed through 20-mesh screen

↓

Dry the residue at 100°C for 1 hour

↓

Compare residual weight to the original wt in terms of percentage

Determination of Sinking Rate

1. Put a piece of feed into a glass beaker
2. Observe the sinking rate and express the results in terms of length or diameter of the feed/second and expressed for 1.5 m depth of water.

Water stability of feeds is greatly influenced by a number of factors. They are

- ☆ Composition of diets
- ☆ Manufacturing processes
- ☆ Binders used.

Composition of Diets

Ingredients which are difficult to grind or to have no binding properties should be kept to a minimum (*e.g.* rice bran, bone meal).

Hydroscopic ingredients such as salt, sugar and molasses absorb water, making the feed moist and crumbly even before being disposal. Generally, stacking products have good binding properties and gelatinization of the starch in the manufacturing process renders the final product more stable.

Manufacturing Processes

Grinding increases the surface area of a feed and thereby permits more space for steam condensation during conditioning processes resulting in harder and more desirable pellets.

Binders

It increases the hardness and water durability of pellets. Binding reduce the void space in the mixtures and thus provide a more compact and durable pellet.

Self Assessment Questions

1. Why aqua feeds are evaluated?
2. Mention the different uses of evaluation in general.
3. Mention the types of evaluation used for feeds.
4. Merits and demerits of physical evaluation.
5. What are the other names used for physical evaluation.
6. Mention 5 quicker methods to evaluate quality of feeds.
7. Water stability of feeds means what?
8. How will you determine water stability of feeds?
9. List out the quality characteristics of good aqua feeds.

Chapter 15

Chemical Evaluation of Feeds and Feed Ingredients

Proximate Analysis: Compositions and Determinations

The compositions of categories in proximate analysis normally considered are:

Water-Water

It is determined by drying a sample in a hot air oven until a constant weight is reached. 5 hours of drying in vacuum oven at 95 – 100°C (or) 2 hr drying at 135°C.

Crude Protein

Essential amino acids, non essential amino acids, free amino acids, amines, nucleic acids

It is determined by kjeldahl procedure in which N_2 content is directly measured using a conversion factor (6.25). [$N_2 \times 6.25$ (100/16)]. Most protein have 16 per cent N in their composition (1mg N = 6.25 mg of protein).

Ether Extract

Triglycerides, Phospholipids, Sterols, Fat-soluble vitamins Miscellaneous lipids (Waxes etc.)

It refers to the fat or lipid content for sample. Using soxhlet apparatus, dry sample is extracted with hot ether. After extraction, the ether is evaporated and weight of the material extracted is determined.

Crude Fibre

Insoluble Polysaccharides (cellulose, chitin, hemicelluloses).

It measures the material remaining in a sample after it has been boiled in weak acid followed by boiling in weak base minus the inorganic residue.

Nitrogen Free Extract

Monosaccharides, Oligosaccharides, Soluble saccharides

Water-soluble vitamins

NFE % on Dry Basis

= 100 – (% moisture + % ash + % fibre + % of fat + % protein) or

= Dry matter – (% ash + % fibre + % of fat + % protein)

All are dry matter basis.

Ash

Essential elements, Non-Essential elements, Toxic elements

It measures inorganic materials that remain after a sample is burned at 600°C. This temperature is sufficient to burn the organic material is a sample.

Chemical Test for Test Diet

Protein Quality

1. Protein pepsin digestibility
2. Protein dispersibility index
3. Urea's activity
4. Available lysine

} Soybean meal (items 1–3)

Protein Pepsin Digestibility

It is used for estimate the digestibility of animal protein feeds. (Digesting a defatted sample for 16 hr in a warm solution of pepsin under constant agitation).

Protein Dispersibility Index (PDI)

This is a method that has been known for nearly 25 years, but gained attention recently. It measures protein solubility after the Soybean meal sample is blended in water at high speed (8500 rpm, 10 minutes): the lower the solubility, the higher the heat treatment. The advantage of this method is that there is a linear decrease in PDI with increase in heat treatment. Value below 45 per cent indicates adequate heating to destroy ANFs. Lower limits to indicate over heating have not yet been established. (A low protein Dispersibility Index indicates more complete destruction of trypsin inhibitors and urease).

Urease Activity

It measures the activity in residual urease in soybean products.

Urease is a heat-labile enzyme and its presence indicates that the soybean meal has not been adequately processed. It is measured by increase in pH due to release of ammonia from urea. The method is relatively simpler to implement, therefore it is used widely in the industry. The limitation of this method is that the rise in pH is not proportional to the extent of heating. Dramatic lowering of urease activity occurs within a short span of time during the heat processing. So, it is difficult to determine whether a meal has been over-heated. The acceptable urease levels were one considered to be 0.05-0.2, *i.e.*: values above 0.2 indicated under heating and values below 0.05 indicated over-heating. However, research has indicated that meals with urease activity levels as high as 0.5 and as low as 0 to be nutritionally as valuable as meals in the 0.05-0.2 range. As a result, urease activity is not considered to be a sensitive indicator of Soybean meal quality, particularly for over processed meals.

Available Lysine

When proteins containing lysine are heated in the presence of reducing sugar, a reaction occurs between the epsilon N_2 of lysine and reducing sugars producing a complex that is undigestible by fish.

Organoleptic Evaluation

Dark, brown colour of the meal with a burnt smell indicates an extremely over-processed soybean meal. However, slight to moderately over processed soybean meal look normal. An extremely low moisture content (<7 per cent) of the meal may also indicate an over processed meal.

Trypsin Inhibitor Analysis

One of the heat labile ANFs in soybean is trypsin inhibitor (TI). While high TI levels indicate under-processed SBM, low or zero TI levels could indicate adequate or over-heated meals. The methods to measure TI levels are difficult to implement in routine feed analysis laboratories, so TI levels are not typically used to assess soybean meal quality.

Protein Solubility in KOH

This method measures protein dissolved in a 0.2 per cent solution of KOH after stirring the soybean meal sample in it for 20 minutes. The lower the solubility, the higher the heat treatment. Solubility levels below 70 per cent indicate excessive heating. The disadvantage of this method is that it is not sensitive enough to detect inadequate heating.

The following table summarizes the advantages and disadvantages of different methods used to verify whether soybean meals have been adequately processed. Among the various methods, PDI seems to be the best because of its sensitivity. It is also a relatively simple method that can be implemented in routine feed analytical laboratories. Until indicative PDI values for over-heating are established, the method can be combined with KOH solubility or urease activity analysis to assess the quality of soybean meal.

Comparison of Methods to Evaluate Soybean Meal Quality

Test	*Observation/Values*		*Pros*	*Cons*
	Under-heated	*Over-heated*		
Organoleptic	—	Colour–Dark Brown; Smell: Burnt	Simple	Subjective
Trypsin Inhibitor (TIU/g)	> 10	<10	Direct measure- of ANF ment over-heating	Difficult to measure; Not sensitive

Contd...

Contd...

Test	Observation/Values		Pros	Cons
	Under-heated	Over-heated		
Urease activity	> 0.2	< 0.05	Relatively simple	Not sensitive to over-heating
KOH solubility (%)	—	<70%	Simple	Not sensitive to under heating
PDI (%)	> 45%	Yet to be established	Simple; Linear relationship between heating and PDI values	Indicative values for over-heating not yet established.

Lipid Quality

Two major concerns with lipids are hydrolytic and oxidative rancidity. Both are undesirable in dietary lipids and in finished feeds.

Hydrolytic Rancidity

It is caused by the hydrolytic or liberation of fatty acids from triglycerides and is detected by elevated levels of the fatty acids in a fat or oil.

Oxidative Rancidity

It is caused by the reaction of O_2 with double bands of unsaturated fatty acids. The products are hydro peroxides – peroxides – aldehydes – ketones. The methods used are:

Peroxide Value (PV)

It measures the initial products of lipid oxidation. Good quality oils usually have PV < 1 milli equivalent.

TBA

It measures another intermediate product of lipid oxidation. TBA (thio barbituric acid) numbers in oils increase with decrease in oxidation. More TBA numbers, less oxidation takes place.

Anisidine Value

It measures the presence of aldehyde rather than intermediate product in a sample. It is useful method for oil, not for feeds because of colour interference caused by chromogens in the diet.

Kries Fest

It is a rapid test which indicates oxidative rancidity when a red colour appears in a sample mixed with phloroglucinol. Red colour indicates presence of aldehydes.

Ash

Fish meal an often tested for ash and NaCl content to meet the specifications required by many fish food industry. 5-6 per cent NaCl in fish meal is undesirable.

Antinutritional Factors and Toxins

Some of the ANFs can be destroyed by heat treatment. Those that cannot be destroyed include gossypol in cotton seed meal; phytic acid in soybean meal and cotton seed meal.

Chemical Score (CS)

The quality of protein in a protein source is decided by the quantity of essential amino acid content present. The essential amino acid content of source is compared with that of a standard protein. The usual standard protein used by the Nutritionists is hen's egg white. CS in calculated as follows:

$$\frac{\text{Limiting amino acid in test protein (g)}}{\text{Limiting amino acid in whole egg protein (g)}} \times 100$$

e.g., Tryptaphan. in egg white: 1.7 per cent; Tryptaphan in sardine:1.2 per cent;

$CS = 1.2/1.7 \times 100 = 70.59$ per cent

Essential Amino Acid Index (EAAI) or Indispensable Amino Acid Index (IAAI)

It is the geometric ratio of the indispensable amino acid in the test protein (TP) divided by the indispensable amino acid or essential amino acid in whole egg protein (WEP).

EAAI Function

☆ To determine deficiency in essential amino acid.

☆ To determine which combination of feedstuffs to meet amino acid requirement of fish.

EAAI – 0.9 (good quality protein); 0.8 (useful); Below 0.7 (Inadequate); Squid meal has 0.9 EAAI and Casein has 0.8.

The EAAI or IAAI is calculated as follows:

$$\sqrt[n]{\frac{\text{aa TP}_1}{\text{AA WEP}} \times \frac{\text{aa TP}_2}{\text{AA WEP}_2} \times \frac{\text{aa TP}_3}{\text{AA WEP}_3} \times \ldots\ldots \frac{\text{aa TP}_n}{\text{AA WEP}_n}}$$

where,

aa TP_1, aaTP_2 = essential amino acid in test protein *i.e.* Feed

AA WEP_1, AAWEP_2 = Essential amino acid in whole Egg protein *i.e.* Fish.

n = No. of Essential amino acid entering into the calculation.

(or)

$$\frac{\text{Arg TP}}{\text{Arg WEP}} + \frac{\text{His TP}}{\text{His WEP}} + \ldots\ldots \frac{\text{Val TP}}{\text{Val WEP}} \times 100$$

New Methods for Determining Nutritional Quality of Fish Meal

D-aspartic Acid Concentration as an Indicator of Temperature Treatment during Drying

Use of D-aspartic acid as an indicator is based on the premise that most amino acids in proteins naturally occur in the biologically usable L-form and that during processing some of them are converted to the D-form (the change from one form to the other is called racemization). The nutritional value of D-amino acids is apparently lower than that of the L-amino acids because of lower efficiency of proteolysis, intestinal absorption and further metabolic processing to convert them into the L-form. Some D-enantiomers may also be nutritionally antagonistic and toxic Aspartic acid is among the most sensitive amino acids to processing inducted racemization, so the author proposes that D-aspartic concentration is an indicator of temperature treatment during drying of fishmeal. The amino acid is measured by means of HPLC and the result is expressed as per cent D-aspartic acid in the total aspartic acid. Several datasets are presented to demonstrate that D-aspartic acid is an effective indicator of fishmeal. Fish meals dried at low temperature had 0.5 – 1.2 per cent D-aspartic acid, while those dried at high temperatures had levels varying from 2.0 to 5.8 per cent.

While further studies are required to validate and standardize the method, D-aspartic acid level seems to be a reliable indicator of fishmeal quality. The main drawback would be the practical difficulties in implementing the method in routine feed analytical laboratories.

Oxysterols Concentration as an Indicator of Lipid Oxidation in Fishmeal

Oxysterols are oxidation products of cholesterol. Apparently, cholesterol is highly sensitive to oxidation when heated or stored in the presence unsaturated fats and under low moisture conditions. Capillary gas chromatography (GC) is used to measure the oxysterols. The results are expressed as mg oxysterols per kg of total lipids. In a preliminary study, two oxysterols were detected in fish meals: 7*b*-hydroxycholesterol and 7-ketocholesterol. Another preliminary study showed that the two oxysterols increased by up to 350 per cent of their initial content after 42 days of shelf storage and then decreased. While the increase indicates the potential use of oxysterols levels as an indicator of lipid oxidation, the subsequent decrease lowers the method's usefulness as a sensitive indicator. Other methods measuring lipid oxidation also suffer from the same drawback and therefore are unreliable. However, the studies have so far been preliminary in nature, so it is too early to comment on the practical application of this method to measure lipid oxidation.

Self Assessment Questions

1. The term "proximates" means what?
2. Mention the factors analysed in proximate composition of feeds.
3. What are all methods available for crude protein estimation?
4. What are all methods available for crude fat estimation?
5. What are all methods available for moisture estimation?
6. What is Ashing?

7. What is exactly crude protein, crude fat, crude fibre, ash, nitrogen free extract?
8. List out the chemical test for test diet.
9. Mention the different types of rancidity.
10. How EAAI is calculated?
11. Peroxide Value, Kries's test, TBA value meant for what?
12. What is meant by chemical score?
13. What is meant by IAAI?

Chapter 16

Biological Evaluation of Feeds and Feed Ingredients

This method does not provide any information on chemical composition, but it provide more accurate information about nutritional value.

Biological evaluation of feed ingredients and finished feeds involve feeding fish and analyzing some aspect of fish performance and or diet digestibility.

Biological evaluation methods can be divided into 3 general categories:

1. **Retention studies or Carcass deposition** in which the deposition of a nutrient in the carcass over a short time in measured. The retention of specific nutrients or energy in the carcass of fish over a specific time period can be a useful way of evaluating availability and balance of amino acid and essential elements. Carcass deposition may also be expressed as apparent retention (AR)

$$AR = \frac{\text{(Carcass nutrient content at end of expt)} - \text{(Carcass nutrient content at start of expt)}}{\text{Nutrient intake during experiment}}$$

2. **Loss studies** in which the various losses of ingested feeds via the faces, urine and gills are measured.
3. **Performance studies** in which some measure of growth are used to evaluate and compare feeds.

In other words, Biological Evaluation methods are organized into 3 groups.

1. General methods used for various nutrients.
2. Methods used for proteins
3. Methods used for energy.

General Methods

Growth

Over a specific time period, growth of groups of fish fed various experimental diets are calculated and compared.

Daily Instantaneous Growth Rate (GW).

$$\text{SGR or GW} = \ln W_1 - \ln W_0 / T.$$

W_1 = Wt. at the end of study

W_0 = Wt. at start of study

T = Time interval in days.

To convert 'GW' to per cent increase in Wt/day (%W/day),

use per cent W/day = $(e^{GW-1}) \times 100$

FCR (Feed Conversion Ratio)

Apparent FCR =

$$\frac{\text{Food given (Supplementary feed + Natural feed)}}{\text{Weight gain/(g)}}$$

FCE

Feed conversion efficiency – It is the reciprocal of FCR and converted in to per cent

$$\frac{\text{Wt gain}}{\text{Food consumed}} \times 100$$

Condition Factor (k)

$$k = \frac{\text{Weight (g)}}{\text{Length (cm)}^3} \times 100$$

The relationship between weight (w) and length (l) can be used to determine condition factors for fish. In farmed fish, condition induces may be used to confirm visual inspections as to whether the fish have a typical conformation or are too big or too thin. The results obtain may predict a change of feeding level or a switch to a feed type with a different nutrient density. (Rain bow trout – 1.3 – 1.6; Atlantic salmon–1.0 – 1.2; Channnel catfish–1.0 – 2; Common carp–1– 2.5).

Methods Used to Evaluate Protein and Amino Acid Quality

PER – Protein Efficiency Ratio–It is measure of wt gain/unit protein fed and is a useful method to compare protein in a single experiment.

PER is calculated as PCR = Wt gain (g)/Protein fed on dry Wt. basin (g)

PCE–final carcass protein – initial carcass protein/protein fed × 100

PCR = Protein gained/protein compound

Net Protein Utilization (NPU) or BV × Digestibility

It is measure of the protein gain during an experimental period/ unit protein absorbed by the fish

(NPU – Final body protein – Initial body protein/Total protein fed × 100).

Apparent NPU is calculated as follows:

$$= \frac{\text{(Protein content of fish at end of expt.)} - \text{(Protein content of fish at start of expt.)}}{\text{Dry protein fed (g)} \times \text{Protein digestibility}}$$

True NPU is calculated as follows:

$$\frac{\begin{pmatrix}\text{Final carcass protein} - \\ \text{Initial carcass protein}\end{pmatrix} - \begin{pmatrix}\text{FC protein of protein free} \\ \text{dietary treatment} - \text{IC protein} \\ \text{of protein free treatment}\end{pmatrix}}{\text{Protein fed} \times \text{Protein digestibility}}$$

Biological Value (BV)

BV measurements are used to determine the percentage of absorbed N_2 retained by a fish by measuring nitrogen excreted during a test period. BV is thus similar to carcass deposition or apparent retention.

1. **Apparent BV** =

$$100 \times \frac{\text{Feed N} - (\text{Faecal } N_2 \text{ Urinary } N_2 + \text{Branchial } N_2)}{\text{Food nitrogen}}$$

True BV =

$$\frac{\begin{matrix}\text{Food N} - (\text{Faecal N} - (\text{Metabolic faecal N}) - \\ (\text{Urinary N} - (\text{Branchial N})\end{matrix}}{\text{Nitrogen fed}}$$

Protein conversion Ratio (PCR) = $\dfrac{\text{Protein gained (g)}}{\text{Protein consumed}}$

Assimilability of protein =

$$\frac{\text{Protein consumed} - \text{Faecal protein}}{\text{Protein consumed}} \times 100$$

Digestibility

Digestibility portion, which has been absorbed, or which has not been recovered in faeces. When digestibility is expressed in per cent, then it is called digestibility coefficient. It is a measure of biological availability of nutrients in the ingredients. Two methods are commonly used to measure digestibility viz. Direct and indirect methods.

Direct Method–Apparent and True Digestibility

It requires quantitative collection of faecal matter. It involves feeding a specific amount of experimental diet and carefully record quantity of feed consumed and faeces produced. The amount of particular nutrient remaining in the faeces is then subtracted from initial quantity of feed. The difference represents the amount of nutrient absorbed by animal. Apparent Digestibility (AD) – Direct method; True Digestibility (TD) – Direct method.

$$TD = \frac{\text{Nutrient in diet} - (N_2 \text{ in faeces} - \text{ Non feed nutrient in faeces})}{\text{Nutrient in diet}} \times 100$$

or

$$A/I = \frac{\text{Absorbed nutrient}}{\text{Intake nutrient}}$$

$$AD = \frac{\text{Nutrient in diet} - \text{Nutrient in faeces}}{\text{Nutrient in diet}} \times 100$$

Indirect Methods

$$AD = 100 - \frac{\%\, Cr_2\, O_3 \text{ in feed}}{\%\, Cr_2\, O_3 \text{ in faeces}} \times \frac{\%\text{ of N in faeces}}{\%\text{ of N in feed}}$$

$$TD = 1 - \frac{\%\, Cr_2\, O_3 \text{ in feed}}{\%\, Cr_2\, O_3 \text{ in faeces}} \times \frac{(\%\text{ of N in Faeces in exptl. diet} - \%\text{ of N in faeces in control diet})}{\%\text{ of N in feed}}$$

It is based on the conc. of an indigestible in the feed and faeces. The marker such as $Cr_2\,O_2$, Fe O, SiO_2 is mixed thoroughly with feed (1 per cent). The feed is then offered to fish for a long periods to get faecal excretion of feed intake. A sample of feed and faeces is analyzed for the nutrient and the marker.

Other Methods Available

1. Tracer studies
2. *In vitro* digestibility

pH Drop Method

As proteolytic enzymes attack the peptide bonds of protein, H_2 is released and pH is reduced. pH reduction is highly and positively correlation with degree of protein digestion.

pH Static Method

To keep digestive enzymes close to optimum pH, pH can be maintained by adding NaOH consumed in proportional to the degree of protein hydrolysis. (Inexpensive technique)

Problem and Sources of Error with Digestibility Measurements

Many problems are associated with the determination of digestibility co-efficient such as size and age of the animal, types of feed processing and processing condition, environmental parameters, interaction with other nutrients, method of collecting the faeces, type of inert material is marker used, teaching of nutrients from feed and faeces.

Methods Used to Evaluate Dietary Energy

Energy values can be determined by measuring energy losses at each stage of digestion and metabolism or by comparative slaughter techniques and by estimation of metabolizable energy and not energy for production.

Self Assessment Questions

1. What is Biological evaluation?
2. Mention the methods available for biological evaluation.
3. What is carcass deposition, Apparent Retention, condition factor?
4. What is meant by digestibility, BV, NPU, PER, PCR, PCE?
5. Mention the methods used to measure digestibility studies.

Chapter 17

Microbiological Evaluation of Feeds and Feed Ingredients

Manufactures should be aware that bacterial contamination is an ever-present threat in fish feed manufacture. In fish feeds, a number of bacterial species are commonly found. A programme of periodic monitoring of ingredients and finished feeds for total bacterial counts, coupled with an aggressive programme of in-plant sanitation and use of commercial antimicrobials in the feed would be prudent practices for all fish feed manufactures. Test include:

1. Total aerobic bacterial count (cfu)
2. *Salmonella*
3. *E. coli*
4. *Staphylococcus*

Chapter 18

Feeding Strategies

There are three distinct feeding phases for fishes (according to age). They are:

I phase – resorption of yolk sac of the hatchlings

II phase – spawn, fry and fingerling stage

III phase – fingerling to table size fish

Food can be offered to fish in excess, apparent satiation and restricted quantity.

Feeding to Excess

Fishes should be fed ad libitum level *i.e.* constant availability to fish. In livestock, feeding to excess is common; but in fish, it is not so. It leads to poor FCR. disintegration of uneaten food cause water quality problem in the culture systems.

Feeding to Apparent Satiation

Feeding of fish can be done with maximum amount of feed that consume. For carnivorous fish, it will give good results.

Feeding to Restricted Amounts

It is other wise called predetermined ration and it is good one.

Practice of Feeding

Practice of feeding in an aquaculture system involves the following 3 stages:

1. Feeding ration (How much feed should be given? Or feed ration or ration size)
2. Feeding frequency (How many times the fish fed in a day ?)
3. Feeding methods (by what methods feeds should be offered?

Ration size

Ration size is defined as the amount of feed/diet time (*i.e.* 24 hrs) made available to fish. Ration size is variable and juvenile fish needs higher ration; grower fish require less ration. Ration size needs to be modified according to the size and age of the fish. The ration size is normally calculated as a per cent of the biomass present

DFR – Daily feed ration is calculated by the following formula:

$$\frac{A \times B \times SR \times RF \times 1\,kg}{1000 \times 100}, \textit{where,}$$

A = Average Body Weight of fishes (g)

B = Number of fish stocked

SR = Survival rate (per cent)

RF = Rate of feeding (per cent)

1000 = to convert to kg.

100 = percentage

A method for determining ration size for the salmonids using body length, past growth rate record and FCR (Piper *et al.*, 1982).

$$\text{Body weight to be fed daily (per cent)} = \frac{FCR \times 3 \times A \times 100}{B}$$

where,

FCR = food conversion ration

A = daily increase in length (cm)

B = length of fish in cm at present time

3 = used because weight has cubic relationship to length

An example of the calculation involved is given below:

Suppose, on 13 April, there are 200 000 fish present in the pond. Their feeding rate was last determined on 1 April, when the fish were 9.07/kg per 1 000 fish (or 1.45 cm in length).

We wish to adjust the feeding rate again now, knowing from past records that at this operating temperature, the average length increase per day (A) will be 0.0075 cm during April and the expected FCR is 1.2.

The new feeding rate is then calculated:

Length, 1 April: 1.45 cm (9.07 kg/1 000)

Plus growth, 13 days × 0.0075 cm (A): 0.0975 cm

= Calculated length today (B), 13 April: 1.5475 (11.03 kg/1000)

Applying the formula, the amount of feed to be given, as a percentage of biomass, is:

$$\% \text{ to feed daily} = \frac{1.2 \times 3 \times 0.0075 \times 100}{1.5475} = 1.7 \text{ per cent}$$

Thus, the weight of feed to apply is:

$$\frac{200000 \times 11.03 \times 1.7}{1000 \times 100} = 37.5 \text{ kg daily, 13 April, for 200000 fish}$$

Advantages

- ☆ Minimal sampling
- ☆ Feeding rates predicted well in advance
- ☆ Reduction in stress, mortality and injury

Disadvantages

- ☆ Availability of long term data
- ☆ Well applicable to intensive system
- ☆ Used for selected species only

Five Basic Rules for Feeding (Piper *et al.*, 1982)

1. For optimum growth and feed behaviour, each feed should ideally be 1 per cent of the body weight. There fore, if the ration for the day is 5 per cent body weight/day, fish needs to fed 5 times, 1 per cent body weight each time (FR = 5 per cent = 5 times of body weight/each time at 1 per cent of body weight)
2. Survival rate is not significantly influenced by feeding frequency once the transition from an endogenous to an exogenous food supply is complete.
3. Higher feeding frequencies reduces starvation and stunting, there by resulting in uniform in size
4. Dry feeds needs to be distributed more frequently than moist feeds.
5. At least 90 per cent of the given feed should be consumed within 15 minutes of feeding time.

Feeding Rate

This is vitally important for efficient aquaculture. Underfeeding can result in loss of production. Overfeeding will cause wastage of expensive feed and is additionally a potential cause of water pollution, a condition resulting in loss of fish or requiring expensive corrective measures. Thus, both overfeeding and underfeeding have serious economic consequences, which affect the viability of the farm.

Sometimes you may read a vague instruction, like 'feed 5 per cent of biomass per day' for a dry feed. This might be applied throughout the growing cycle. This would almost certainly result in near starvation in the early stages, gross overfeeding, and water quality problems later. Feeding rates should not stay steady throughout the whole of the growing cycle to market size. They must be modified according to the size and age of the fish or shrimp, and to the water conditions.

The quantity of feed to be given to a pond or cage each day should normally be based on a percentage of the biomass present (total weight of animals). Thus, if a pond contains 10 000 fish weighing 10 g on average and the recommended feeding rate (see later) is stated to be 7 per cent per day, the amount of feed to be given daily is:

$$\frac{10000 \times 10 \times 7}{100} = 7000 \text{ g (7kg)/day}$$

The percentage of biomass to be fed is not a fixed amount. It should decrease as the animals grow, to reflect their decreasing metabolic rate. Thus, the ratio of weight of feed per day to animal weight (biomass) is high at the start of the growing period and lowers towards the time when the animals reach marketable size.

Applying a feeding rate accurately depends on an accurate estimate of average fish weight and of the numbers of fish in the production unit (pond, cage, etc.) (survival). Average weight can be obtained directly from weighing samples or by measuring the length of the fish, where an accurate length/weight relationship has been established. Accurate record keeping is essential not only to aid efficient feeding now, but also to enable you to examine the effect of past actions and to help you predict the result of planned actions during the next growing cycle.

Feeding tables have been constructed for various aquatic species. Manufacturers of compound feeds always give a feeding guide for their products. The tables for those species, such as trout, which are reared under highly intensive conditions tend to be more elaborate and reliable because they are based on many decades of accurate observation and measurement. They usually specify feed type and size as well as the daily feeding rate. Similarly, elaborate tables will become available for other species as more becomes known about the most efficient ways to present their feed. Meanwhile, simple feeding guides for some species are available.

It is emphasized that the feeding rates given in tables must not be applied without reference to other factors. Feed should be reduced in quantity or omitted during times of low temperature and based on operational experience in a specific location and environment, increased when growth rates are predicted to be highest. Daily feeding rates must also be based on observation of the animals during feeding. At this time feeding activity, water quality (colour), presence of old feed, etc., must be assessed. All feeding tables are merely a guide which, if applied with careful judgment, will markedly improve economic viability. However, if applied rigorously without complementary assessment of conditions, they can result in disaster.

As stated before, feeding rates should be decreased as the animals grow. Feeding a steady percentage of biomass throughout the growing cycle usually results in underfeeding when the fish are small, depressing growth and survival rates and overfeeding when the fish are larger. The total effect is depressed growth rate, poor survival and poor apparent feed conversion ratio. The more often feeding rates are adjusted, the more efficiently will feed be utilized. Ideally, they should be adjusted daily, but this would require too much paper work and would confuse those who do the feeding. Also, it is only possible to adjust the feeding rate accurately when the biomass can be estimated by measurement or when it can be accurately predicted from past experience. This is where accurate, long-term records are so important.

In practice, feeding rates are adjusted weekly or twice-monthly for salmonids and catfish, based on estimates of biomass, together with knowledge of environmental conditions (mainly temperature). Feeding tables for other species are at present less refined and feeding rates are adjusted less frequently. For example, using a feeding table given for carp, the amount of feed to be given daily (at 20–23°C) begins at 9 per cent for animals of less than 5 g in size. It changes to 7 per cent for animals of 5–20 g and to 6 per cent for those of 20–50 g average weight. Further reductions, to 5 per cent, 4 per cent, and 3 per cent, respectively for animals weighing 50–100 g, 100–300 g, and 300–1000 g are recommended in that table. Obviously, changes in feeding rate will be less frequent when this type of table is used than is possible with the more detailed tables, for salmonids. In practice, more frequent adjustments can be made to feeding rates even when simple feeding tables are used. In the example just mentioned, feeding rates at 20–23°C could be adjusted more frequently in the following way:

Feeding Tables

Animal Size (g)	*% of Biomass to be Fed per Day*	
	Recommended in Table	*Adjusted Actual Rate*
5	7	7.0
10		6.7
15		6.3

Contd...

Table 1.1–*Contd...*

Animal Size (g)	*% of Biomass to be Fed per Day*	
	Recommended in Table	*Adjusted Actual Rate*
20	6	6.0
30		5.7
40		5.3
50		
	5	5.0
60		4.8
70		4.6
80		4.4
90		4.2
100		4.0

Feeding rate tables for trout, salmon, channel catfish, common carp, tilapia and marine shrimp are provided.

Fish Size (g)	*Crumble and Pellet Size*	*Amount of Feed*	*(%Body Weight/Biomass) Per Day*			
		At: 7°C	*9°C*	*11° C*	*13°C*	*15°C*
0.38	No. 1	3.4	3.9	4.8	5.8	6.4
0.77	No. 1	3.3	3.8	4.7	5.6	6.1
1.43	No. 2	3.0	3.6	4.5	5.1	5.8
2.5	No. 2	2.8	3.2	4.0	4.9	5.1
5.0	No. 3	2.6	3.0	3.8	4.5	4.7
7.7	Nos. 3-4	2.3	2.8	3.6	3.9	4.1
11.1	No. 4	2.0	2.4	2.9	3.2	3.8
25.0	2.4 mm	1.7	1.9	2.1	2.6	3.2
33.3	2.4 mm	1.6	1.8	1.9	2.2	2.9
50.0	3.4 mm	1.4	1.6	1.8	2.1	2.5
66.7	3.4 mm	1.3	1.5	1.7	2.0	2.4
100.0	4.8 mm	1.2	1.4	1.6	1.8	2.0
200.0	4.8 mm	1.1	1.3	1.5	1.7	1.9
500.0	6.4 mm	0.9	1.0	1.1	1.3	1.6

Source: NRC, 1981 (simplified and adapted).

Feeding Table for Pacific (Coho) Salmon Fed Oregon Moist Pellet 1/

Fish Size (g)	*Amount of Feed*	*(% Body Weight/Biomass) Per Day*			
	At: 40°F	*45°F*	*50° F*	*55°F*	*60°F*
	(4.4°C)	*(7.2°C)*	*(10.0°C)*	*(12.8°C)*	*(15.5°C)*
Below 0.76	3.2	5.1	7.3	8.9	10.5
1.5	2.3	3.4	5.2	6.4	7.5
2.5	1.9	2.9	4.4	5.3	6.3
4.0	1.6	2.6	3.8	4.6	5.5
5.0	1.5	2.5	3.5	4.3	5.1
7.0	1.2	2.2	3.0	3.8	4.5
10.1	1.0	1.9	2.8	3.5	4.1
13.4	0.9	1.7	2.5	3.2	3.8
20.2	0.8	1.4	2.1	2.8	3.3
25.2	0.7	1.2	1.9	2.5	3.0
34.9	0.5	0.9	1.5	2.0	2.5
Above 41.3	0.4	0.8	1.3	1.8	2.3

Source: Piper *et al.*, 1982 (simplified and adapted).

1/: Amounts are greater than for a corresponding dry diet because the dry matter content of the feed is less.

Examples of Feeding Rates for Trout Fed Commercial Feed

Animal Size (g)	*Feeding Rate (per cent Biomass/Day)*		
	At: 5°C	*13°C*	*19°C*
0-18	3.3	6.3	9.3
23–40	1.1	2.0	3.0
180 plus	0.5	1.0	1.6

Source: Sales literature, Nippon Haigo Shiryo K.K. (Japan).

Feeding Rate for Tilapia with Commercial Pellets

Fish Size (g)	*Feeding Rate (per cent Biomass/Day)*
<10	9-7
10-40	8-6
40-100	7-5
>100	5-3

Source: Sales literature, President Enterprises Corporation, (Taiwan).

Feeding Rate for Tilapia Fed 25 per cent Protein Feed in Monoculture at 24°C

Fish Size (g)	*Amount of Feed*	
	(g/fish/day)	*(per cent of biomass)*
5-10	0.5	10-5
10-20	0.8	8-4
20-50	1.6	8-3.2
50-70	2.0	4-2.9
70-100	2.4	3.4-2.4
100-150	2.7	2.7-1.8
150-200	3.0	2.0-1.5
200-300	3.7	1.9-1.2
300-400	4.5	1.5-1.1
400-500	5.2	1.3-1.0
500-600	6.0	1.2-1.0

Source: Marek, 1975 (adapted).

Feeding Rate for Tilapia (*T. nilotica*) in Tanks and Cages at 27-31°C Fed a 46 per cent Protein Commercial Fish Feed

Fish Size (g)	*% of Biomass to be Fed Per Day*
Up to 5	30 reducing to 20
5-20	14 reducing to 12
20-40	7 reducing to 6.5
40-100	6 reducing to 4.5
100-200	4 reducing to 2
200-300	1.8 reducing to 1.5

Source: Pullin and Lowe-McConnell, 1982.

Feeding rate for Tilapia with Commercial Pellets

Fish Size	*Feeding Rate (per cent Biomass/Day)*
<25 g	8-6 per cent
>25 g	4-3 per cent

Source: Sales literature, Tai Roun Products Co. (Taiwan).

Typical Spring-Summer-Autumn Feeding Schedule for Channel Catfish in Ponds Based on Stocking Rates of 5 000-7500/ha and a 36 per cent Crude Protein Diet 1/

Date	*Water Temperature*		*Fish Size*	*% Body Weight to Feed*[2/]
	(°F)	*(°C)*	*(g)*	
Ar 15	68	20.0	18	2.0
Apr 30	72	22.2	27	2.5
May 15	78	25.6	50	2.8
May 30	80	26.7	73	3.0
Jun 15	83	28.3	95	3.0
Jun 30	84	28.9	127	3.0
Jul 15	85	29.4	159	2.8
Jul 30	85	29.4	191	2.5
Aug 15	86	30.0	272	2.2
Aug 30	86	30.0	341	1.8
Sep 15	83	28.3	404	1.6
Sep 30	79	26.1	459	1.4
Oct 15	73	22.8	499	1.1

Source: Piper *et al.*, 1982.

1/: This table illustrates the combined effects of animal size and temperature on feeding rate. As temperature increases feeding rate also increases but, as the animals grow larger, metabolic rate decreases and lower feeding rates are recommended. 2/Fed six times per week.

Commercial Feeding Programme for Channel Catfish Fed Dry Feeds 1/

Fish Size	*Feed Type*	*Feeding Amount (% Biomass Per Day)*	*Feeding Frequency*
Swim-up	Trout Chow size 00	Ad lib on water surface	3-10/Day
Up to 2.5 cm	Trout Chow No. 1	Ad lib on water surface	4/Day
Up to 3.8 cm	Trout Chow Nos. 2 or 3	3 per cent	4/Day
3.8-7.6 cm	Trout Chow Nos. 3 or 4	3 per cent	3/Day
7.6-12.7 cm	Trout Chow No. 4 or Catfish Starter	3 per cent	3/Day
12.7-17.8 cm	Catfish grower	32-45°F: 0.5 per cent	2/Week
		45-55°F: 1.0 per cent	4/Week
17.8 cm-market	Catfish chow (in ponds)	55-65°F 2.0 per cent	6/Week
	or	65-75°F 3.0 per cent	6/Week
	Catfish cage chow (in cages)	75-85°F 4-6 per cent	7/Week
		85-95°F 3-4 per cent	5-6/Week
Broodstock	Catfish breeder chow	32-45°F 0.5 per cent	4/Week
		45-55°F 1.0 per cent	5/Week
		55-65°F 2.0 per cent	6/Week
		65-75°F 3.0 per cent	6/Week

Source: Ralston Purina, 1974

1/: Like previous Table, this feeding chart illustrates the combined effect of increasing animal size and seasonal water temperatures on recommended feeding rates.

Feeding Rates for Channel Catfish Fed Floating Feed

Fish Size		*per cent Biomass to be Fed Daily at Various Temperatures (°C)*					
(mm)	*(g)*	*15*	*18*	*21*	*24*	*27*	*30 plus*
76	4.4	2.0	2.5	3.1	3.5	4.0	4.4
102	10.5	1.7	2.2	2.7	3.1	3.5	3.9
127	20.5	1.5	2.0	2.4	2.7	3.1	3.4
152	35.4	1.4	1.8	2.1	2.5	2.8	3.1
178	56.2	1.2	1.6	1.9	2.2	2.5	2.8
203	83.9	1.1	1.4	1.7	2.0	2.3	2.5
254	163.9	0.9	1.2	1.4	1.7	1.9	2.1
305	283.2	0.8	1.0	1.2	1.4	1.5	1.7
356	449.7	0.6	0.8	1.0	1.1	1.3	1.4
381	553.1	0.6	0.7	0.9	1.0	1.1	1.3

Source: Foltz, 1982 (simplified).

Feeding Rate of Channel Catfish with Commercial Feed

Animal Size		*Feeding Rate (per cent Biomass/Day)*			
		60°F	*70°F*	*75-85°F*	*>90°F*
(Inches)	*(g)*	*15.5°C*	*21°C*	*24-29°C*	*>32°C*
6	27	1	3	4	2
9	91	1	3	3	2
12	227	1	3	3	1.5
15	472	<1	1.4	1.4	0.7
18	863	0.8	-	-	-

Source: Sales literature, Western Grain Company, Alabama, USA

Feeding Rate for Marine Shrimp (*Penaeus monodon*)

Animal Age/Size	*Feed Type*	*Feed Rate (per cent Biomass/Day)*
Up to P30	PL No. 1	25-20
P30–0.6 g	PL No. 2	20-16
0.6 g–5 g	Starter No. 1	16-9

Contd...

Contd...

Animal Age/Size	*Feed Type*	*Feed Rate (per cent Biomass/Day)*
5–10 g	Starter No. 2	9-7
10–20 g	Grower	7-5
20 g–market size	Finisher	5-3

Source: Sales literature, Hanaqua Feed Corporation, Taiwan, 1984.

Experimental Feeding Guide for Carp

(per cent Body Weight/Biomass Per Day)

	Pellet Size:					
	1.5mm	*1.5mm*	*2.7mm*	*4mm*	*5mm*	*5mm*
Temperature	*Animal wt:*					
(°C)	*>5*	*5-20*	*20-50*	*50-100*	*100-300*	*300-1000*
<17	6	5	4	3	2	1.5
17-20	7	6	5	4	3	2
20-23	9	7	6	5	4	3
23-26	12	10	8	6	5	4
>26	19	12	11	8	6	5

Source: A. Coche (pers. comm.)

Feeding Frequency

The most effective method of feeding with respect to location, time of day and frequency varies from species to species. Its cost effectiveness depends also on other factors such as the availability of feeding labour or automatic feeders, size of pond or tank, cost of labour and the personal preference of the farm manager based on observations and results.

Feeding frequency means number of times and periods/diet time (24 hours) made available to fish. Returning to the subject of feeding frequency, the following sub-sections summarize information on a species by species basis.

Salmon and Trout

In common with other very young fry of fish and shrimp, very frequent feeding is most effective for young salmonids. For swim-up fry of salmonids, the daily feed ration is split up into very small quantities fed as often as 20-24 times per day, either manually or automatically. Sometimes, a 24-hour lighting regime is used for the first few days to encourage the fry to take dry feed. Feeding frequency is gradually reduced to 1-3 times per day as the fish grow. Rainbow trout start to take food about twenty-one days after hatching when reared at 10° C. Most hatcheries feed at ½ to 1 hourly intervals during an 8-hour day, reducing this to three times per day. After the fish are about 5 inches long (23 g) feeding frequency is reduced to twice per day. Brood fish are fed only once per day. Feeding frequencies quoted by Piper *et al.* (1982) for coho salmon, autumn chinook salmon and rainbow trout are given in following Table.

Salmonids are frequently reared in tanks or cages so feed distribution is not a problem. Many different types of automatic feeders are utilized for salmonids.

Suggested Feeding Frequencies for Salmonids

Species	*Fish Size (g)*								
	0.3	*0.45*	*0.61*	*0.91*	*1.82*	*3.6*	*6.1*	*15.1*	*>45.1*
	No. of Feeding Times Per Day								
Coho salmon	9	8	7	6	5	3	3		
Autumn Chinook salmon	8	8	8	6	5	4	3		
Rainbow trout	8	8	6	6	5	4	4	3	2

Source: Piper *et al.*, 1982.

Salmonids tend to feed to satiation and then do not eat again until most of the meal has left the stomach. Once past the fry stage therefore, a feeding frequency of 1 or 2 times per day is sufficient.

Catfish

The information presented here refers to channel catfish; it is reasonable to assume that it is applicable to other species of catfish.

Channel catfish fry begin to feed 5-10 days after hatching, when the yolk sack reserves are used up. As with salmonids, swim-up fry are best fed many times per day. One commercial feed manufacturer recommends 8-10 feeds per day, reducing quickly to 6 per day by the time the fish are about 2.5 cm long. Feeding frequency is further reduced to 3 times per day when the fish reach 7.6 cm in length. Juvenile catfish grow best with two feeds per day, one at mid-morning and one late in the afternoon, seven days per week.

Feeding frequency in ponds depends on water temperature. At 13-29°C, feeding 6-7 times per week is recommended. At times of particularly high or low temperatures, less frequent (4 or 5 times per week) feeding is suggested. Feeding 6, instead of 7 days per week is said to encourage the fish to consume any surplus feed in the pond and lessen the chance of over-feeding. Catfish in cages should be fed daily. There is some evidence that feeding catfish twice per day, especially under raceway conditions, results in a faster growth rate.

Tilapia

In the wild, tilapia feed more or less continuously throughout the day. Manual feeding, several times per day is best for intensively grown tilapia, in cages or raceways for example.

Automatic feeding can be employed and the blower type of feeder is said to distribute feed more adequately for tilapia. Tilapia fry should be fed at least 4 times per day, preferably 8 times per day, in daylight hours. In an experiment with 9 mm total length fry of Oreochromis aureus, New *et al.* (1984) showed that survival improved with continuous (mechanical) feeding compared to either five or three manual feeds per day. Feeding rates can be less frequent, 4 or 5 times per day for fingerlings. Adult tilapia thrive best on 2-3 feeds per day.

Carp

Again, common carp (and probably Chinese and Indian carps) thrive best on frequent feeds. Jauncey (1982) reports that one researcher found that optimum feed utilization by common carp (at 40 g size) was achieved when the feed was split into nine equal feeds. The best feeding frequency can only be assessed on an individual farm basis, determined by the cost of repetitive feeding operations. From the biological and nutritional point of view, it would appear best to feed as frequently as possible.

Other Fish Species

Specific feeding frequency recommendations for the other cultured species of fish covered in this manual are not available. It is therefore recommended that until more information on the optimum conditions becomes available those frequencies found best for catfish and tilapia should be applied. A golden rule would be, when in doubt, to feed as frequently as economics allow. There has been a report, however, that groupers gave best biomass increase and good FCR, fed 'trash' fish to satiation in cages, when fed every second day, as compared to other frequencies varying from once every 5 days to three times per day. However, this result was apparently caused by poorer survival at the higher feeding frequencies. Best growth rate was, again in this case, achieved by feeding two or three times per day.

Shrimp and Prawns

There is a good deal of controversy about the optimum time and feeding frequency for marine shrimp and freshwater prawns. Some species burrow during the day and feed most actively at night. Others feed in the shallower parts of the pond, but avoid these areas in daylight when temperatures are highest. For these species, it would seem best to feed in the late afternoon or early evening. Most farmers feed once, or at the most twice per day, usually first thing in the morning and last thing in the afternoon.

Shrimp do not consume all of the feed presented at once, unlike most fish. This fact has led to much discussion and research on methods of binding shrimp feeds to prevent wastage and the loss of water-soluble nutrients. Some commercial feeds are extremely water stable (>24 hours), but may not be so palatable as softer feeds. The apparent need to produce such well-bound diets has, in part, been caused by the reluctance to feed shrimp and prawn ponds more frequently than once a day, even though labour is often available and otherwise unoccupied. Feed presented in smaller quantities more frequently would not need to be so efficiently bound and should therefore be cheaper.

In Taiwan, many intensive farmers (usually farms are run by a family enterprise, so there is always someone on site) feed tiger shrimp (*Penaeus monodon*) four to six times per day, the feeds being evenly spaced over the whole 24 hour period.

Feeding shrimp and prawns as frequently as possible, spreading the daily ration between those feeds, would be the most effective technique. It certainly pays off with young post-larvae, as it does with fish fry. Shrimp larvae do not thrive at all unless maintained in a conditions where food (live food or artificial feed with neutral buoyancy) is constantly available.

- ☆ Carps – 5 to 3 times/day–day feeding
- ☆ Milk fish – 8 times/day–day feeding
- ☆ Cat fish – 2 times/day–day feeding
- ☆ Tilapia – 1 time/day–day feeding
- ☆ Murrel – 1 time/day–day feeding
- ☆ Shrimps – 5 times/day–day and night feeding

  6am – 20 per cent–Day time

  10am – 10 per cent–Day time

  2 pm – 10 per cent–Day time

  6 pm – 30 per cent–night time

  10 pm – 30 per cent–night time
- ☆ Prawn – 3 to 5 times/day–day and night feeding

Normally, finfishes require day time feeding (6am – 6pm); where as shellfishes requires day and night feeding (6 am – 6 am).

Feeding Methods

Cultured fish may be fed by one of two ways: I) by broadcasting or hand feeding or manual feeding and II) by feeders (mechanized)

Dry pellets, granules and crumbles for fish could be broadcasted. For small ponds (less than 0.50ha), broadcasting is quite adequate. For larger ponds, broadcasting should be supplemented by distribution in slightly interior areas using a small boat. Moist or semi moist should not be broadcasted; but, kept in earthern pot or plastic trays; Placed in the peripheral areas of the pond bottom. The adequate number of feeding trays required is 30 to 40/ha to create maximum opportunities for fish

In narrow or small ponds for fish, feed should be spread evenly around the perimeter. For larger ponds, other methods have to be used to give adequate distribution, especially for species which are territorial in nature. These methods include feeding from a boat, feed blowers towed by a tractor, and (in the U.S.A.) distribution by aeroplane for very large ponds. Boats are, of course, essential for the feeding of moored cages which are not connected to the shore by a walk-way.

Manual Feeding–Advantages

- ☆ Behaviour of fish is regularly observed

Manual Feeding–Disadvantages

- ☆ Labour intensive
- ☆ More time consuming
- ☆ Limited in application on large farms

Feeders

Feeders are of 2 types, namely:

1. Demand feeders and
2. Non-Demand feeders (Automatic feeders)

Demand Feeders

It has several types, namely, Response feeders (for larger fish), Plate feeders (for small fish), Demand feeders with sensor.

Non-Demand Feeders

In this, a predetermined quantity of feed at predetermined time intervals is dispensed to pond. The quantity of feed dispensed is determined by the power of the motor or by pressure of air stream and by size of the pellet. Non – demand feeders are of the following types, namely, Drop feeders, Auger feeders, Disc feeders, Pneumatic feeders

Drop Feeders

It uses belt dispensers or rotating disc for dispensing the feed. Rotating disc permits more controlled delivery of feed. No regulator is provided. As a consequence, do not dispense precisely quantity. It is mainly used to fed crumbles, starter feed to juvenile stock.

Auger

As the name implies, these feeders dispense feed by auger. The quantity of disposal is regulated by period of operation. It broadcast up to a distance of 4-5m in the pond.

Disc

It is used for pelleted feeds (more than 3 mm) over a wide area. Disc is mounted at set distance below the hopper. When it is activated, feeds falls on disc and broadcasted over the desired area.

Pneumatic

It utilize either high or low pressure air stream for dispensing the feed. The quantity of feed delivered in to path of air stream by regulator. High pressure feeders are driven by compressed air (stationary) an low pressure feeders are driven by fan commonly used.

Feeding Devices

The feeding of fish and shrimp is done by hand in most farms and there are advantages in doing so. The main one is that it enables the operator to inspect his stock regularly and to judge whether they are eating properly. It also enables him to check the other parameters of the pond/tank/cage at the same time.

There are, however, a number of mechanical aids to hand feeding and many types of automatic feeders on the market. Automatic feeders are particularly appropriate to intensive systems and the feeding of nursery fry tanks which require frequent, small doses of feed.

Automatic feeders are available for dry diets. Moist diets are difficult to dispense automatically because of their texture. The exact operational details of each feeder are not illustrated, only the principle involved. Details of devices for fry feeding are given in another FAO publication (Berka, 1973). Commercially available feeders are marketed by aquaculture equipment supply companies in each country.

Some feeders, particularly demand feeders, are relatively easy to construct using simple materials like oil drums or plastic containers.

Mobile Devices for Dry Feeds

Feed does not necessarily have to be carried round a pond. It can be pushed round in a wheelbarrow or, if the ponds are large or many, and the bunds are wide and strong enough to take a vehicle, the feed can be towed in a truck or a tractor driven trailer. The feed can then be shoveled or thrown into the pond by the operators.

Feed can be more efficiently distributed in this manner with mechanical help. Nearly all equipment of this type depends upon a blowing device powered by the truck or tractor engine. Feed is released into the turbo blower by the operator who controls the time (and therefore the amount of feed) and the direction in which it is ejected. This then is still a form, albeit mechanically aided, of manual feeding. Hoppers in this type of equipment can contain up to 3-4 tons of feed at a time and blowers will distribute feed over an area of up to 6 × 3 m on each occasion or up to a distance of 20 m from the pond bank.

Blower feeders are obviously designed for very large farm units and are not appropriate for small-scale aquaculture: they are mentioned here for completeness.

Similarly, while feed is often transported by boat and fed by hand or by shovel, devices have been used in large farms to aid this operation. These include mobile boat mounted blowers, as above, and longitudinal slots in the bottom of boats through which different amounts of feed can be released by operating a lever.

Stationery Devices for Dry Feeds

These devices can be grouped into a number of categories. Some require mains or battery electrical power. Some rely on waterpower, others on the weight of the feed and the action of the feeding fish.

Electrically Powered Feeders

These fall into two broad categories–those, which operate mechanically, and those, which employ compressed air.

In both cases the control devices are electrical. The time and duration (thus the amount of feed) can be pre-set by the operator using an electrical timer. This may be mains or battery driven and it may operate a single feeder or a whole bank of feeders. Some sophisticated feeders are controlled not only by timers but also by sensors which detect when certain environmental factors are correct, such as temperature or light intensity.

Compressed Air Feeders

Basically, though there are many variants and patented examples of compressed air feeders, most are based on the same principle. A compressor supplies air to one or a number of feeders.

The supply of air to each feeder is normally shut off. Each feeder has a supply of feed in a hopper mounted above a feed distribution pipe, in turn placed over the tank or pond. Feed is allowed to fall by gravity from the hopper into the distribution pipe. It ceases to flow when the orifice of the hopper becomes blocked by the fallen feed. A blast of air is introduced into the distribution pipe by the release of a valve controlled by a timer and the feed is ejected with considerable force. The amount of feed ejected on each occasion depends on the diameter of the distribution pipe and the hopper outlet and, principally, by the length of time that the blast of air is allowed to pass through the distribution pipe.

Mechanical Feeders

The operation of these types of feeders, which are also controlled by timers, depends on electro-magnets or electric motors. The principles of operation are best described by the following series of diagrams:

Type 1

Movement of the slug 'A' is controlled by an electro-magnet. The space 'B' governs the amount of feed released at each movement of 'A'.

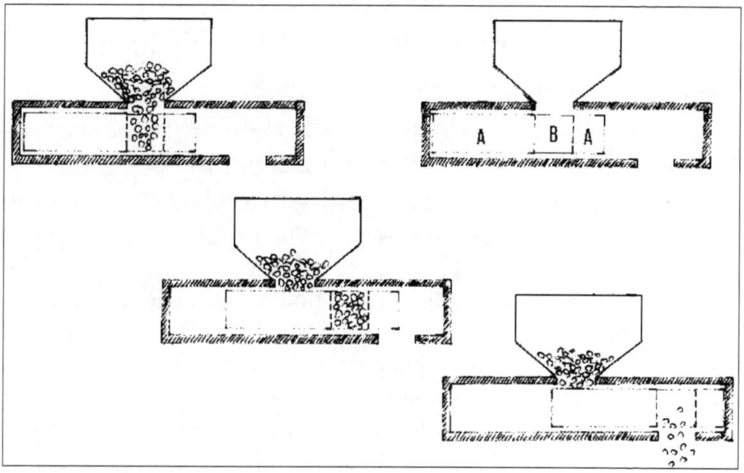

Type 2

Here the feed trough consists of two parts, one inside the other. The movement of the inner one is controlled by an electro-magnet.

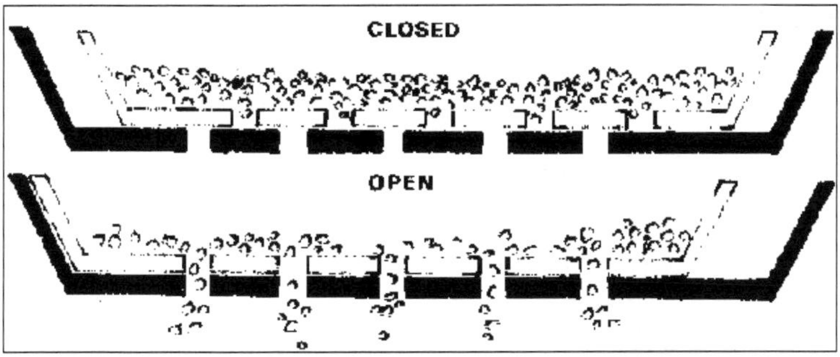

When the holes in the two parts of the feeder coincide, the feed falls through.

Type 3

In this version, the feed falls from the hopper on to a disc which is rotated by an electrical motor at intervals to eject a portion of feed. The motor also releases the feed from the hopper on to the disc by operating a valve. The feed can either be released directionally using the guide shield 'B' or, if the latter is removed, throughout a 360 angle.

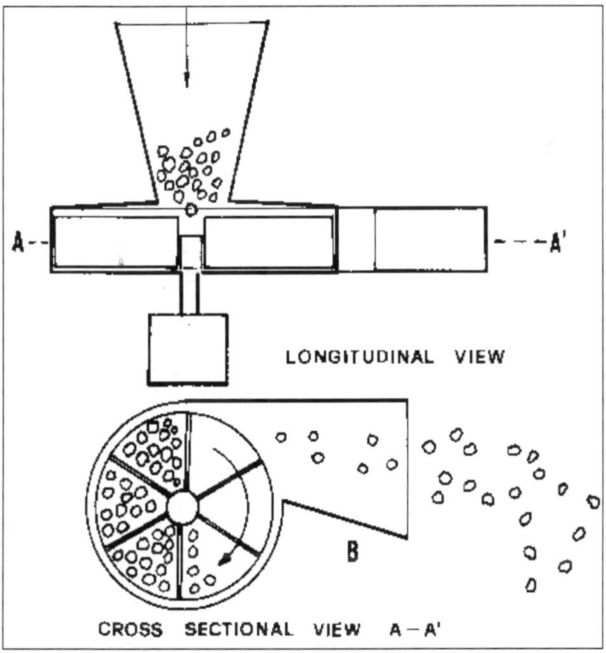

Type 4

In type 4, an endless screw mechanism transfers the feed from the hopper to the outlet. The amount of feed released depends on the number of revolutions of the motor drive screw, which is controlled by a timer as is the periodicity of feeding.

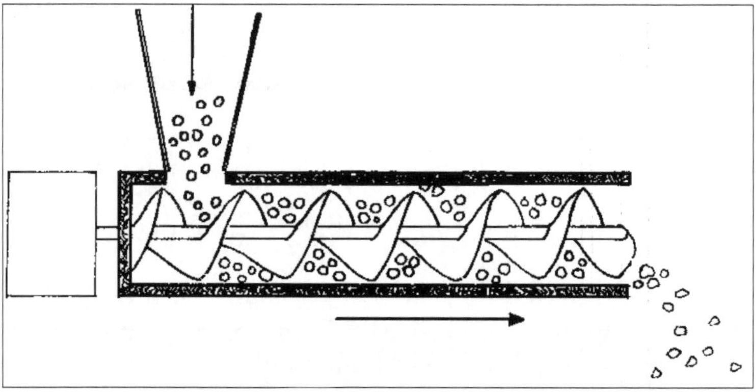

Type 5

Type 5 is similar to type 4 except that a blower is added, which distributes the feed over a greaterdistance.

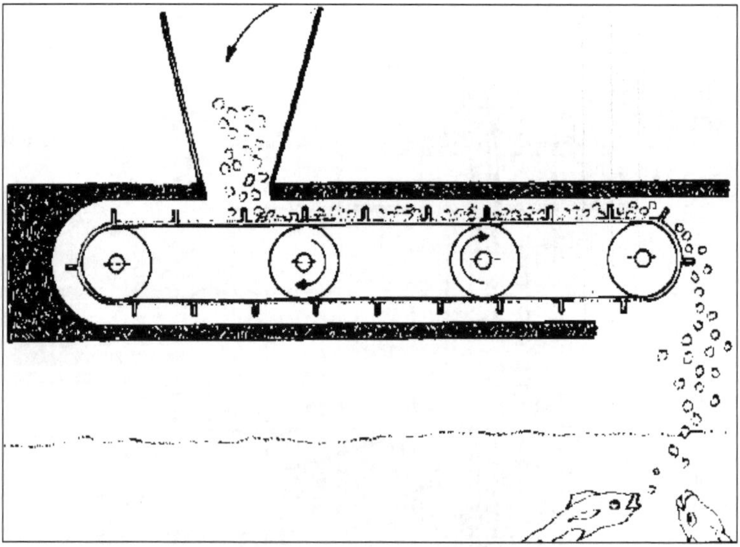

Type 6

In type 6, the feed is delivered on a conveyor belt driven, at selected intervals, by a motor controlled by a time switch.

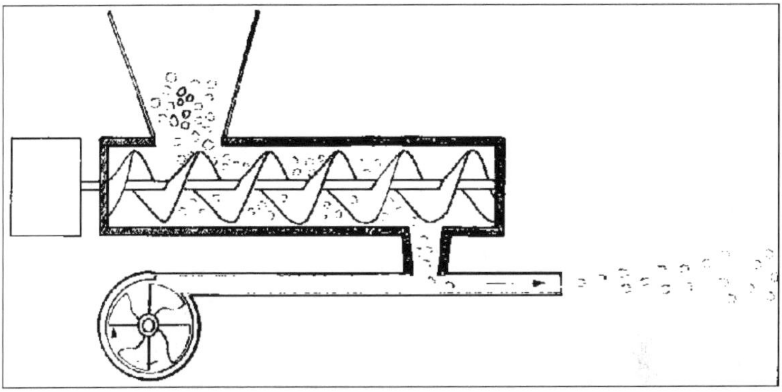

Type 7

In type 7, a series of spikes on a revolving spindle overturn a row of feed containers in turn. The frequency depends on the speed of revolution of the spindle.

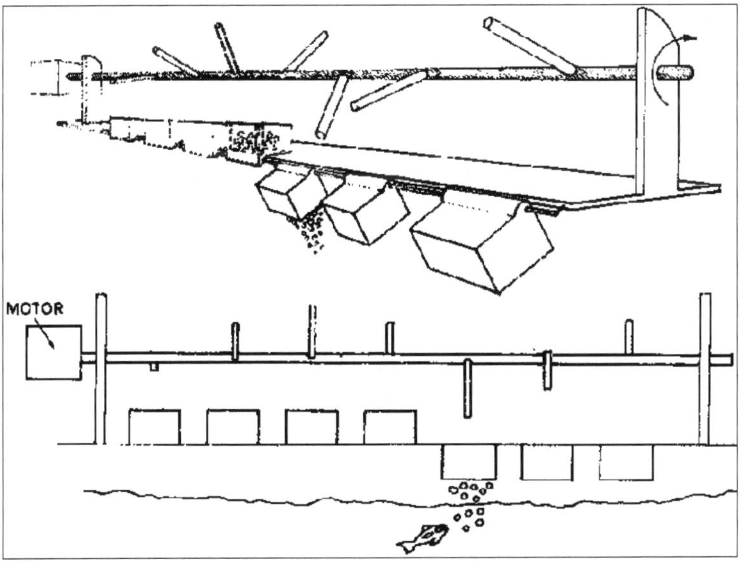

There are many commercial varieties of the feeders whose principles were described above.

Demand Feeders

There are also many different varieties of demand feeders but their general principle is the same. Some species of fish learn very rapidly to use demand feeders but they are usually unsuitable for

small fish which are unable to operate them. The following diagram illustrates the principle.

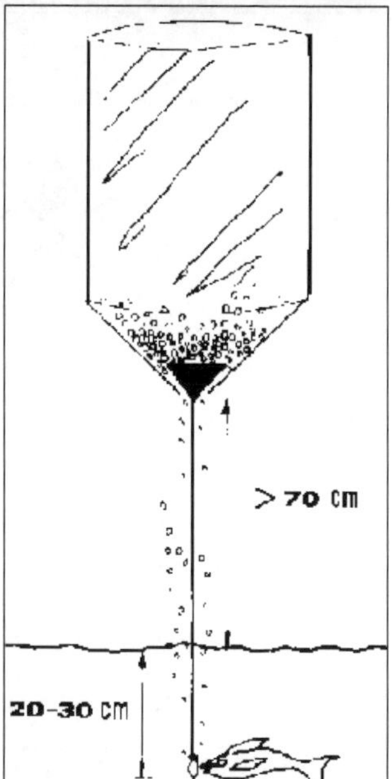

In the demand type of feeder illustrated above, the fish touch the rod connected to a plug or plate in the bottom of the feed hopper. This plug normally closes the hopper so that feed does not fall out. When moved by the movement of the bait rod, a small quantity of feed is released. The quantity of feed released on each occasion can be controlled by the shape and design of the plug. The plug is usually ball shaped or an inverted cone. These feeders can easily be 'home-made' (Hepher and Prugenin, 1981 and Meriwether, 1986).

Another type of feeder has advantages over the normal demand feeder; it relies on the WEIGHT of food consumed. Instead of the bait rod in the example shown above, there is a rod with a feed tray on the end. As the weight of feed on the tray decreases, more feed is released from the hopper.

Water Controlled Feeder

The operation of these feeders is similar to those diagrammed in this section Types 4, except that the motive power is water instead of electricity. Either a water wheel is used or water is allowed to run into a container which empties on a syphon system, like a lavatory cistern, every time it becomes full. As the container empties (or the wheel moves) it triggers a valve on a feed hopper which releases a controlled amount of feed.

Feeding Devices for Wet or Moist Feeds

The non-manual distribution of high moisture feeds is much more difficult than that of dry feeds because of the formers' stickyness. However, the principles involved in some dry feeders can be adapted for use with moist feeds. The examples given in this section Types 4 and 6 can be modified for this purpose but normally hoppers will need to be re-designed to prevent the feed sticking together.

In Japan, where small 'trash' fish is often used for feeding large aquaculture cages, feeding is done in the following way. A feeding boat moors alongside the cage. Alongside the feeding boat itself is a barge full of 'trash' fish. A suction hose is put into the barge and, using a pump on the feeding boat, the feed is sucked up into it. From there the feed is transferred to the centre of the cage using water pumped through a pipe mounted on a boom (see following page).

This type of feed and other moist feeds can also be transferred to large cages and ponds more effectively than it can be fed by hand, by using a mechanical 'thrower'. This is simply a centrifugal fan into which the feed is dropped.

Another type of feeder used for dispensing moist minced feed is based on forcing the feed through a horizontally mounted die plate through the use of a heavy weight.

The weight type of minced feed dispenser runs continously; the system does not lend itself to operation at fixed intervals.

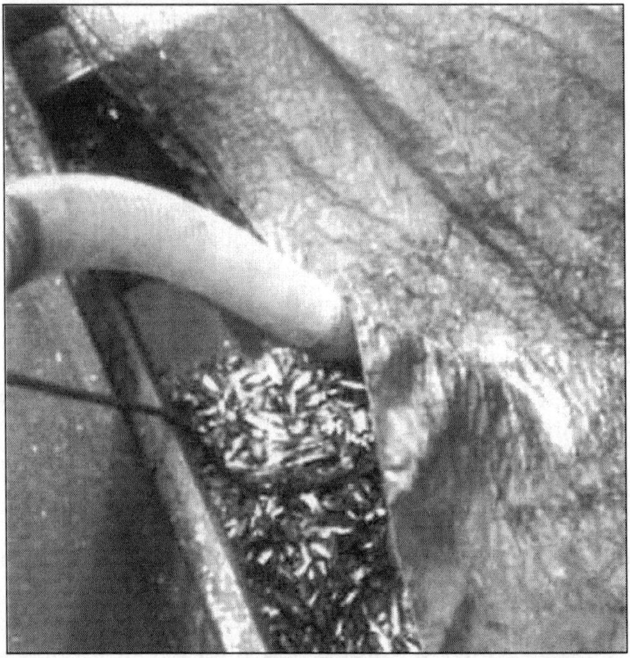

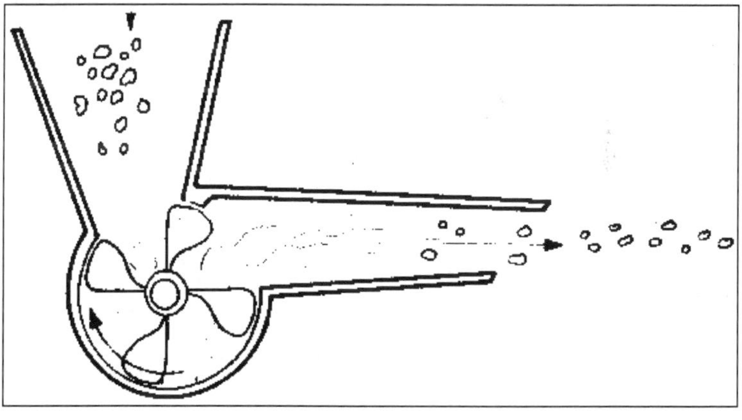

Flowing water is used to transfer freshly made moist pellets in another feeding technique. In this case the feed is made, in the predetermined quantity for each cage, in a mixer/extruder mounted on board a boat. The mixer/extruder is driven mechanically or hydraulically from the boat engine. The extruded feed falls into a trough where a supply of water, provided from a small on-board pump, washes it through a pipe into one or more cages:

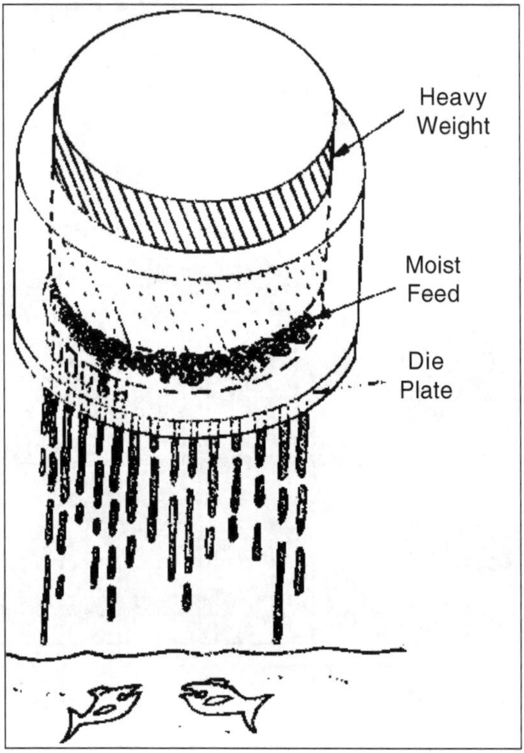

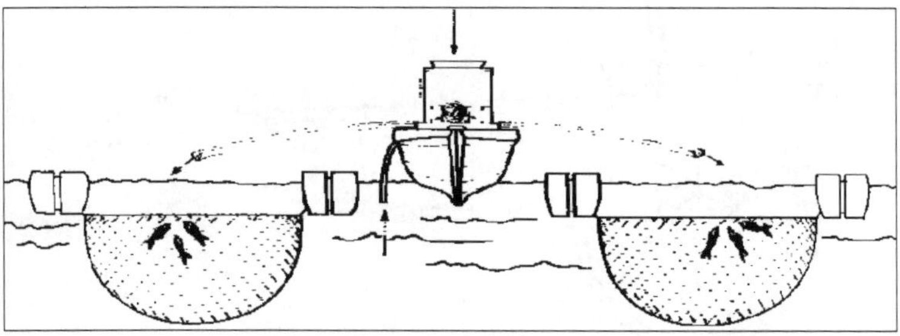

Other Devices

Floating feeds are sometimes put within a floating collar so that they do not float away all over the pond or tank. In this way the feeding activity of the fish can be concentrated in one place. Less feed is wasted and it is easier for the operator to observe the feeding behaviour of the fish.

Biomass Assessment

Calculation of the amount of feed to present daily must be based on an assessment of the biomass (total weight of fish or shrimp) in each pond or enclosure.

Regular, accurate, data for average animal size must be obtained through weekly, bi-weekly or, at the worst, monthly sampling of the rearing unit. In tanks or cages, it is easier to take representative samples, but in ponds it is common to get a biased picture, particularly when a species with uneven growth rate is stocked, such as freshwater prawns. Care must be exercised to take samples in several parts of the pond, not only at feeding points where the larger or more active individuals may congregate. Samples may be taken by seine, cast net or lift net.

If accurate length/weight relationships for the species have been pre-determined under the environmental conditions being used, length measurements are a more accurate means of monitoring growth rate. This is particularly true of crustacea which hold uneven quantities of water under their carapace. Measuring length can be a rapid process with a skilled operator and is less stressful to the animals than trying to determine weight by a standardized technique. Fish length can either be total or, to avoid inaccuracies due to damage to the tail fin, is more accurately measured to the anterior end of the caudal fork. Care must be taken that the method of measurement corresponds with that used when the length/weight ratio was determined. Similarly total shrimp length, because of frequent rostrum damage, is less reliable than measuring from the posterior of the eye orbit to the tip of the telson.

For a pond of 5 000 m^2, stocked at $10/m^2$, at least five samples should be obtained at each sampling time and 50 animals from each sample measured. Average weight can either be calculated directly from the total weight of the sample obtained by weighing or by referring the average of the measured lengths to a length/weight ratio.

An assessment of survival is also necessary to calculate feeding rate effectively. This can be illustrated as follows. If a pond is stocked with 50 000 animals and at a sampling date the average weight is 10 g and the feeding rate to be applied is 3 per cent of the biomass/day, the amount of feed would be:

$$\frac{50000 \times 10 \times 7}{100} = 15000 \text{ g (15kg)/day}$$

if it is assumed that all the animals originally stocked are still present. If however, there has been a 20 per cent mortality up to the 10 g size, the correct amount of feed should be:

$$\frac{50000 \times (100 - 20) \times 10 \times 3}{100 \times 100} = 12000 \text{ g (12 kg)/day}$$

Besides saving 20 per cent of daily feed costs in this case, an accurate assessment of survival, as well as growth rate, would prevent possible water quality deterioration caused by over-feeding.

Although a good estimate of survival aids an effective feeding programme it is often extremely difficult to achieve. In small cages and tanks it is often possible to make accurate visual observations or to count all the animals during their transfer to another tank or cage. This is normally impossible or impractical in pond culture, except when stock transfer from one pond to another takes place for other reasons. In this case the numbers present can be calculated by taking the total weight of stock and dividing by the average animal weight obtained from samples. Visual observation is also impossible in ponds. In my view, no-one has yet devised a satisfactory way of assessing the stock in aquaculture ponds, which is why this important subject is rarely mentioned.

In practice, most farm managers apply an arbitrary survival factor based on the number of days since stocking. This factor is derived from knowledge of previous culture cycles on the same farm or elsewhere in similar circumstances, modified by observation of actual mortalities, knowledge of water quality or disease problems, etc. The factor derived depends on accurate measurement of the number of animals originally stocked and the numbers harvested in each cycle for its accuracy. Thus, the importance of careful farm records becomes obvious.

If those records show, for example, that 50 per cent of the animals stocked normally reach market size (barring individual accidents) on the specific farm, it would be reasonable to assume that future growing cycles would show the same survival rate. Similarly, if there is no transfer of animals within the growing period, the highest mortality rate probably occurs after initial stocking (due to the handling stress and the ease of damage and of predation on young animals). Thus, if a 50 per cent mortality is known to occur normally between stocking and harvesting in a 16 week growing cycle, for example, it would be reasonable to assume that 20 per cent of the losses occur within the first 4 week period with a further 10 per cent loss occurring during every 4 week period after that. Thus, assessments of biomass in this example at two weekly intervals of a pond stocked with 50 000 animals would be based on multiplying the average animal weight, obtained by measurement of samples by the following number of animals:

Weeks 1 and 2 $\quad 50000\ \frac{100}{100} = 50000$

Weeks 3 and 4 $\quad 50000\ \frac{90}{100} = 45000$

Weeks 5 and 6 $\quad 50000\ \frac{80}{100} = 40000$

Weeks 7 and 8 $\quad 50000\ \frac{75}{100} = 37{,}500$

Weeks 9 and 10 $\quad 50000\ \frac{70}{100} = 35{,}000$

Weeks 11 and 12 $\quad 50000\ \frac{65}{100} = 32{,}500$

Weeks 13 and 14 $\quad 50000\ \frac{60}{100} = 30000$

Weeks 15 and 16 $\quad 50000\ \frac{55}{100} = 27500$

Thus, to continue this example, if the average weight of prawns at the beginning of week 13 (12 weeks after stocking) was 20 g and the percentage of biomass to be applied was 5 per cent, the daily amount of feed would be:

$$\frac{30000 \times 20 \times 5}{100} = 30000 \text{ g (30 kg)/day}$$

Clearly assessment of biomass, particularly in ponds, depends partly on accurate sampling and size measurement, but also very much on the manager's judgement based on an accurate record of past experience modified by the history of the particular batch being cultured.

Hand Feeding	*Mechanized Feeding*
Behaviour of fish is regularly observed	Observed only during sampling
Labour intensive	Less labour
More time consuming	Less time consuming
Limited in application on large farms	Any larger size fish farm can be used.
Sensitive to the changing feed requirement of fish	Not sensitive to the feed requirement of fish
Extensive, modified extensive and Semi intensive cost of production/ kg of fish is high	Intensive and Super intensive cost of production/kg of fish is less

Self Assessment Questions

1. What are three distinct feeding phases for fishes?
2. What are different ways of offering food to fishes?
3. What are the three stages of practice of feeding in an aquaculture system?
4. What are ration size, feeding frequency and feeding methods?
5. How to calculate Daily feed ration for carps, shrimps and scampi?
6. What are basic rules of feeding fish?
7. Collect different feeding table's available commercially important fishes.
8. What are the different feeding methods for feeding fish?
9. Compare manual and mechanized feeding.
10. What are types of demand feeders?
11. Compare demand and non-demand feeders.

Chapter 19
Aqua Feed Processing

The term "Feed manufacturing or feed processing" is concerned with the physical transformation of a written formulation into a compounded "edible" diet. Aqua feed processing are of 2 types, namely, (*i*) Dry feed manufacturing and (*ii*) Wet feed manufacturing.

The major steps in dry feed manufacturing process are:

1. Procurement of quality ingredients and additives
2. Storing in warehouse after proper labeling/marketing
3. Preserving labile ingredients (Eg: oils, vitamins in low temp. (*i.e.* cool place).
4. Accurate weighing of the ingredients and additives in proportion required for the select formula.
5. Preparing vitamin and minerals premixes.
6. Grinding of ungrouped materials – Dry ingredients (Hammer mill, attrition mill, Roller mill, cutters) and Wet ingredients – (Wet grinder)
7. Micro grinding
8. Screening
9. Mixing (Horizontal mixer, Turbine mixer, Vertical mixer, Liquid mixtures).

10. Pelleting/Extrusion
11. Cooling/drying (Vertical cooler dryer and Horizontal cooler dryer)
12. Crumbling
13. Screening particle segregation
14. Bagging
15. Storage

Preparation of Feeds

There are two methods are employed in feed preparation. They are, Pelletisation (steaming done separately); Expanded extrusion (Steaming and pelleting takes place simultaneously). In Pelletisation, dry and crumbles are (Sinking alone) produced and compression takes place. Where as in expanded extrusion, Floating, sinking, moist dry feed; expansion takes place.

Pelletization	*Extrusion*
1. Steaming done separately	Steaming and pelleting takes place simultaneously
2. Compression	Expansion
3. Sinking, moist feed	Floating, sinking, moist
4. Generate 2-5 per cent fines in handling of feeds	1–2 per cent
5. Little prolonged storage time	Prolonged storage time
6. Less water stability	High water stability (12 hr. without 24 hrs with binder)
7. Less reduction in microbial load 1,20,000 cfu/g	High reduction in microbial load 10,000 cfu/g.

Grinding

The role of grinding play a major role in feed processing. The important roles are:

1. It reduces particle size.
2. It facilitates the destruction of the heat labile anti-nutritional factors invariably present.

3. It improves nutrient digestibility by increasing the surface area of the feed particles.
4. It improves the mixing property of individual feed ingredients, and also increases the bulk density of the feedstuff.
5. It increases the surface area of ingredients.
6. It removes moisture due to aeration and thereby facilitating mixing, pelleting and digestibility.

For dry ingredients, hammer, attrition, roller, cutters; wet ingredients (wet grinder).

Hammer mills are most efficient to grind, dry low- fat ingredients, plate mills are not suitable for aqua feeds as they are increase of grinding the particles finely. Grinding efficiency of mill depends on number of hammers, their size, arrangement, sharpness, the speed of rotation, the HP of the motor, size of screen mesh used, type of material being ground.

Grinders/Mills for Dry Ingredients

It is used to reduce the size of ingredients so that they are suitable for mixing into a feed. Four types; (*i*) Attrition or plate mills; (*ii*) Hammer mills; (*iii*) Roller mills; (*iv*) cutters. It is used for grinding feed ingredients. Generally, grinding improves feed digestibility, acceptability, mixing and pellatability.

Different types of grinders for dry ingredients are

Hammer Mills (Aqua Industry)

These are impact grinders with swinging or stationary bars forcing feed ingredients against a circular screen or solid serrated section called striking plate. Material in held in the chamber until it is reduced to the size of the openings in the screen. Grinding efficient in dry and low fat feed ingredients. In hammer mills, the grinding chamber consists of series of non-moving or swimming hammers are attached to a rotar. The hammer breaks up the incoming material which is often forced through a steel screen. The steel screens are available with different hole sizes depending on the desired particle size. A cross-sectional view of a hammer mill is shown below:

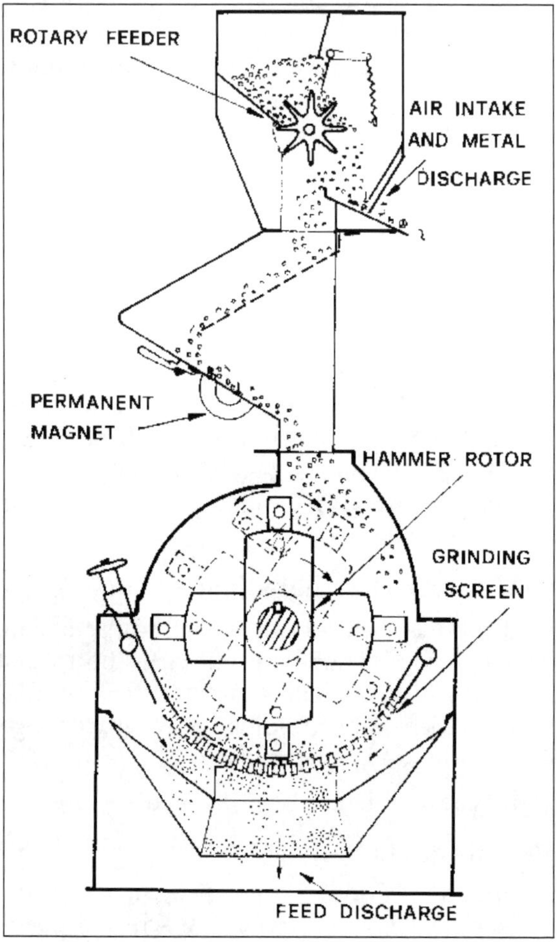

Attrition Mill (Not Used in Aqua Industry)

It consists of two discs rotating in opposite direction. When one disc is rotated, the other stands stationary. The assembly is used for shredding and defibering. Ingredients with liquid which form clumps are grinded in this machine.

Roller Mills

It is a combination of cutting, attrition and crushing. The rollers rotate in a predetermined spaced with a predetermined distance between two rolls. It is economic and but good only for fairly dry and low fat ingredients.

Cutters

It reduces dry solid particles by shearing knife edger against a striking plate. It consists of a rotating shaft with four attached parallel knives and a screen occupying one fourth of the 360 degree rotation. It is best used to crack whole grains.

Grinders/Mincers for Wet Ingredients

It is used to reduce particle size of wet ingredients. The principle of wet feed grinder is the same as that of the domestic mincer. The ingredients fall into a continuous spiral or screw, which is motor drivers. The screw forces the feed against a knife which is rotating against a fixed die plate, through which the feed is forced. Die plates are available with a range of hole sizes depending on the coarseness of the final product desired. Often feeds are minced twice with successively smaller die plates to produce a better result. The product is extruded in spaghetti – or noodle like form.

Advantages:

- ☆ Able to grind many tons of product per hour.
- ☆ Largest mincers are able to tackle both as well as flesh and also frozen blocks.

Premix Preparation

It is prepared in order to facilitate uniform dispersion of vitamin, minerals and additives and it is required in small quantities. Each ingredient must be finally ground (<200mm). Premix is then blended with the feed mix at 30–50 kg/tonne of feed) and mix is ready for use in the final feed mix.

Mixing

The objective of mixing is to obtain a homogeneous dispersion of nutrients and additives so that every unit weight of feed fed to the fish has the same nutritive value.

Generally, dry ingredients are mixed first followed by liquid ingredients. Mixing can be done either a batch or continues process. The types of mixers used vary and it includes

- Horizontal mixtures (continuous ribbon mixtures and non continuous ribbon mixtures) – Dry mixers
- Vertical mixtures–Dry mixers
- Liquid mixtures (or) blenders–wet mixers
- Turbine mixers–Dry mixers

Horizontal types of mixer are preferable for aquaculture feeds than vertical. Bowl shaped mixers with paddles are the most suitable for moist feeds. Carnivorous fish require high oil lipid and it could be sprayed on to the pellets after pelleting. Heat sensitive vitamins, enzymes and other additives can also be sprayed after cooling the pellets. Chemical binders if used should be dissolved in cold or hot water and the solution is added to the feed mix.

Dry Mixers

Two types viz. Horizontal and Vertical–Efficient mixing in the key to good feed production.

Horizontal Mixers: Two types

Continuous Ribbon Mixers

The continuous or "twin-spiral" mixer consists of a horizontal stationary half cylinder with revolving helical ribbons placed on a central shaft so as to the other as the shaft and ribbon rotate inside.

Non-continuous Mixers

Non continuous or interrupted ribbons are similar to the continuous ribbon mixture except that short sections called paddles or ploughs are spaced in a spiral round the mixer shaft. It is more suitable for mixing liquids with dry solids.

It consists of a series of paddles or metal ribbon blades mounted on a horizontal rotor within a semi-circular trough. The blades move the material from one end of the mixer to the other, tumbling, it as it goes. These mixers usually discharge the mixed product from the bottom, using the same mixer blade action.

Another type of horizontal mixer has a bowl shaped or feat bottomed container in which a series of paddles are mounted on a spindle drivers by a motor mounted either above or below the mixer. Horizontal mixers can also be used for mixing moist feeds.

Vertical Mixers

It consists of vertical screws, which elevate the ingredients to the top of the mixer where they fall by gravity to the bottom, to be mixed and re-elevated. These are cylindrical, cone or hopper shaped containers with single or double screw located vertically through the center. The screw operates at speeds of 100 to 200 rpm and vertically, conveys incoming materials from the bottom end like a screw conveyer, to the top where they are scatted and fall by gravity.

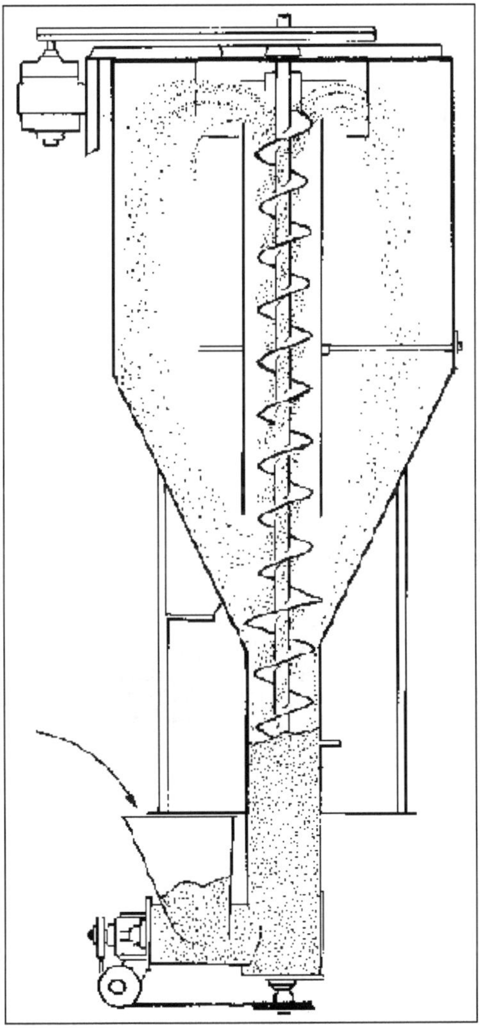

Wet Mixers

Turbine Mixers

Simple bowl or circular mixers, all are the most suitable types for mixing wet ingredients or mixtures of wet and dry materials. They can also be used for mixing a few kilograms or of dealing with several tons of ingredients. The larger types of discharge the mixed product from the bottom. This is done for 10-12 minutes. It can also be loaded from the top.

Liquid Mixers

It consists of a horizontal tub or cylinder with a number of wires or paddles equally spaced around a shaft which revolves inside. Sometimes, the shaft is hollow and liquids are forced through holes in the paddles in a spray affect. Speed is 400 to 1200 rpm. Used for mixing liquid with dry ingredients.

Combined Mixer-Mincer for Moist Aquaculture Feed Production

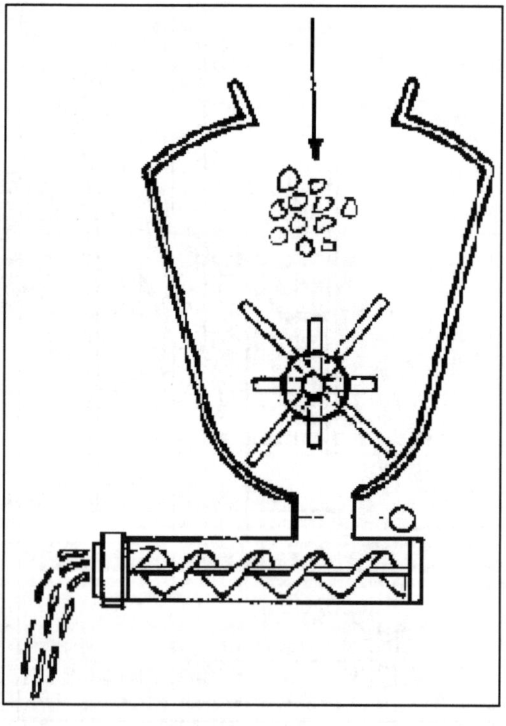

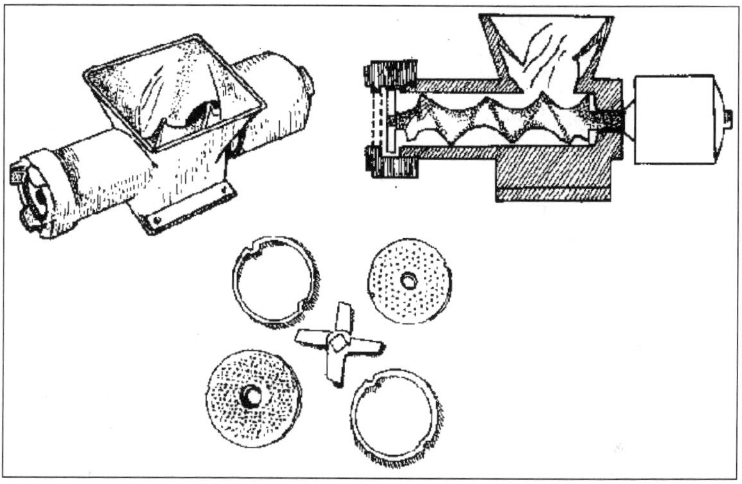

Extruder/Pelletizer

The transformation of a soft, often dusty feed into a hard pellet is accomplished by compression, extrusion and adhesion. Pelletisation is done in two types is, extruder and pelletizer.

Extruder

Principle

In the typical cooking extrusion process, moisture, thermal energy and mechanical energy are added to dry ingredients in order to cook the carbohydrates, sterilize the ingredients and to force the mass through the die where it is cut to size. As the mass exits the die, expansion will likely occur because of the pressure differential between the inside and the outside of the extruder barrel. If the moisture inside of the extruder barrel is heated above 100°C, then there will be a rapid transformation of the moisture to water vapor as it exits the die. This rapid flash will cause the extrudate to expand.

It consists of an extra barrel fitted with a die plate and a shaft conveyor, which is connected to a high speed motor. The feed mixture is fed into an extruder for by proper arrangement of water/steam injection facility. The extruder operators at high pressure (14 – 98 kg/cm^2) and steam pressured (5-7 kg/cm^2) injection. The temperature of the materials rises to 110° C to 130° C for a short time and cooks the food, gelatinizing the starch present in the feed mixture.

Expanded extrusion: Expansion is takes place during process. Extruder operates at high press (14 – 98kg/cm^{2}and steam pressure at 7 kg/cm^2 injection. Water and steam are applied to the feed mix in the conditioning unit to raise the moisture from 10 – 12 per cent to 25 – 30 per cent and feed is conveyed into a pressure – sealed cylinder. Steam injection (5.7 kg/cm^2) increases gelatinsation of starch in the ingredients. The extruded feed strands are cut to pre-determined length out side of the die plate by a rotating knife. Heat sensitive additives can be sprayed on the pellets after extrusion.

Extruder consists of barrel fitted with die plate and staff conveyer, which is connected to high speed motor.

A typical flowsheet for a feedmill utilizing a extrusion-cooking system is shown on next page.

Pelletizer (like noodles or semia or vermicelli making machine)

Used in prawn feed production. Consists of a pair of rollers and a die, which is driven by a high speed motor. It works with a

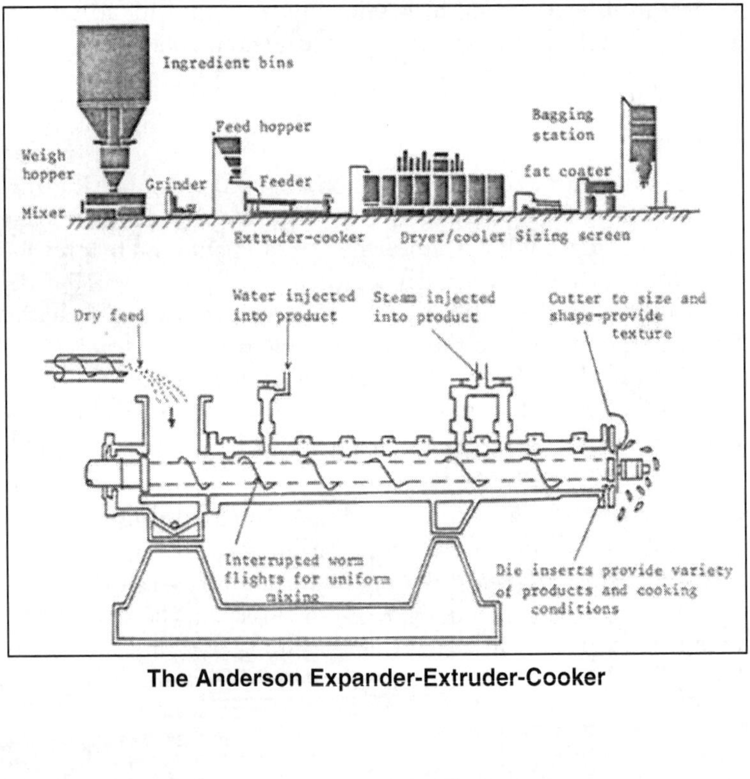

The Anderson Expander-Extruder-Cooker

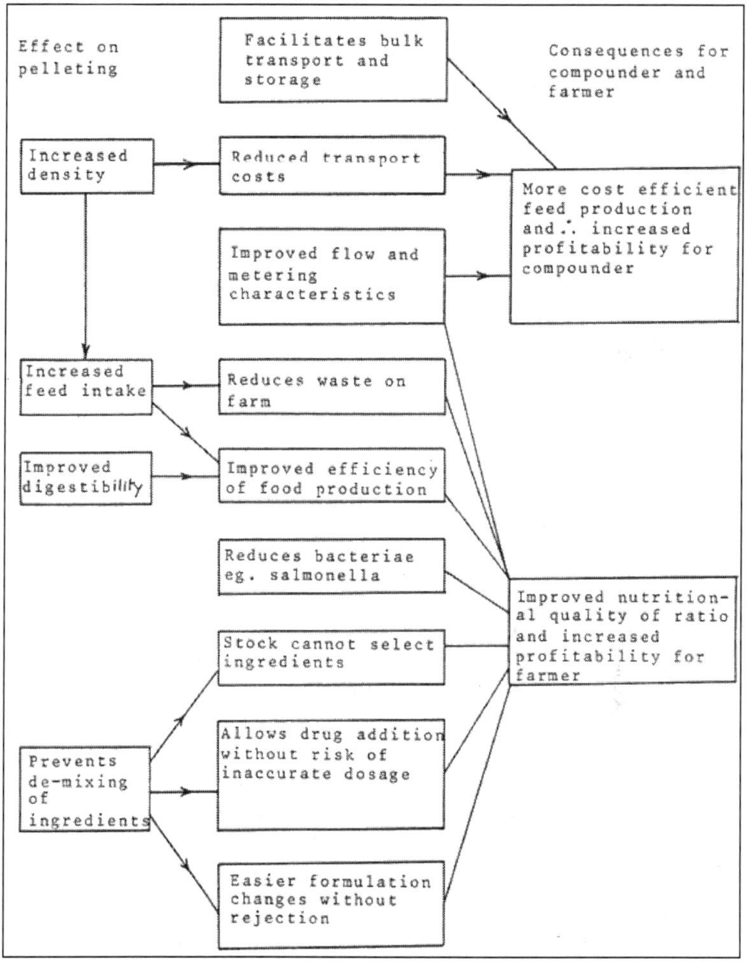

Typical Flow Sheet for a Animal Feed Manufacturing Plant Utilizing a Extrusion-Cooking System (*Source*: Horn, 1979; Williams, 1986)

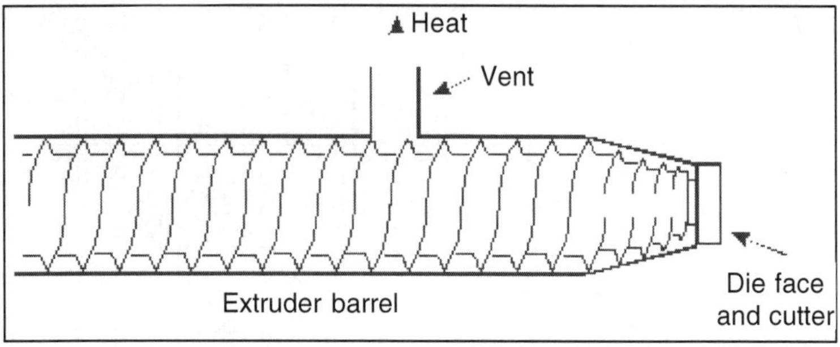

Typical Single Screw Extruder Barrel w/vent

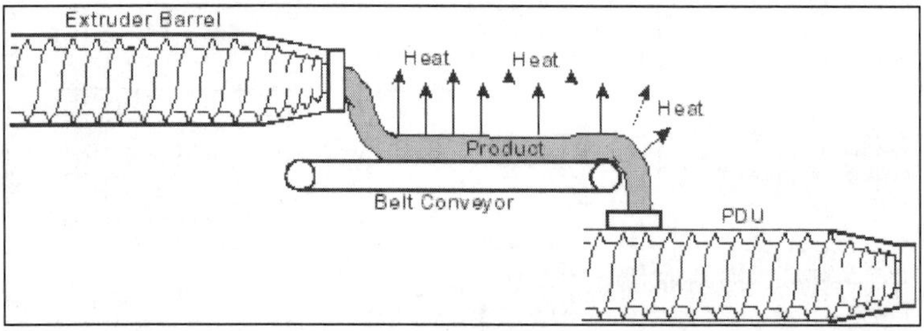

Typical Single Screw Cooking Extruder in Conjuction with a PDU

high pressure (42 – 1800 kg/cm^2) between rollers and the die, steam 0.5 – 3.5 kg/cm^2 and moderate temperature 75 – 95°C. Wet pelletizer works with high moisture levels (30 per cent).

Pelleters

The process of pelleting consists of forcing, a soft feed through holes in a metal die plate to form compacted pelletets which are then cut to a pre-determined size. Presence of heat causes some gelatinization of raw stanch resulting in adhesion. Moist pelleters now have a conditioning unit mounted above them where liquids such as water and molasses are can be added to improve pelletability. The water is some times added in the form of steam, which softens the feed and partially gelatinized, the starch content of the ingredients, resulting in firmer (more water stable) pellets (moisture provides lubrication for compression and extrusion). Passing a feed mixture through conditioning chamber where 4 – 6 per cent water as steam may be added. From the conditioner, the feed fall into the center of the pelleter itself. In the pelleter, two or more rollers and feed ploughs push the material through the holes of the die plate. Within about 20 seconds of the entering the pelleter, feed attains moisture of 15 –16 per cent at 80 – 90°C. During subsequent compression and extrusion, friction further increases feed temp to nearby 92 °C. Pellets discharged on to a screen beet of a horizontal funnel drier and air cooled within 10 min. and dried to moisture below 10 per cent. Usually the die plate itself rotates and on its outerside, stationary knives cut the pellets to a pre-set length. Die hole sizes usually range from 2mm upto 9.5mm; During subsequent compression and extrusion, friction further increases feed temp to nearby 92 °C. Pellets discharged on to a screen beet of a horizontal funnel drier and air cooled within 10 min. and dried to moisture below 10 per cent The thickness of the die plate helps to determine the compactness and stability of the pellet.

Pellet dies can be upto about out 90mm thick. The arrangement of a typical pelleting plant are supply bin, pellet mill, cooler pellet crumble, sifter and collectors.

3 basic types of pellets are made in the aqua industry. Compressed pellets extruded dry pellets; Semi moist extruded pellets.

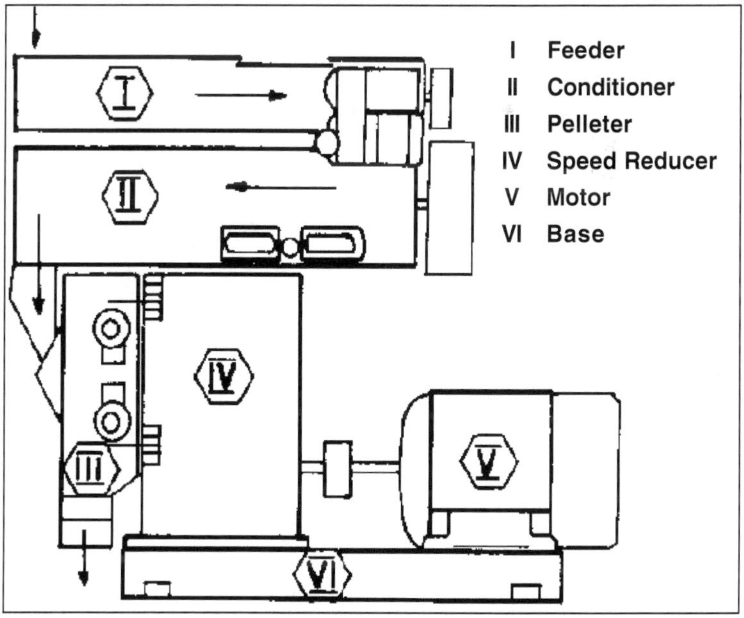

Parts of a Conventional Pellet Mill

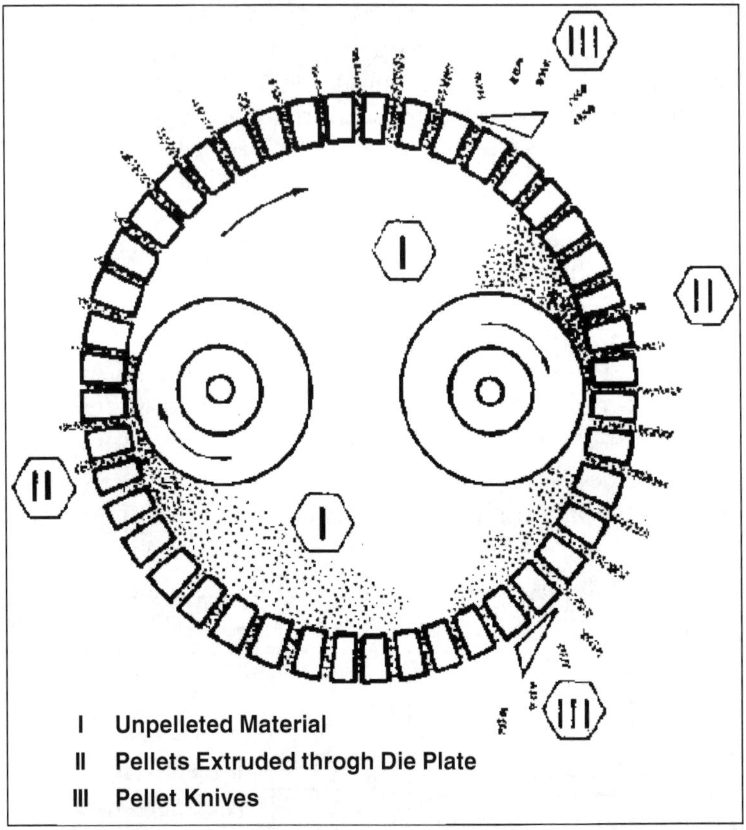

Die Plate from a Commercial Sized Pelleting Machine

Die hole sizes usually range from 2 mm up to 9.5 mm. The thickness of the die plate helps to determine the compactness and stability of the pellet. Pellet dies can be up to about 90 mm thick.

The advantages of the extrusion pelleting over conventional steam pelleting can be summarized as follows:

Extrusion or expansion pelleting is a moist heat process whereby the preground and blended dry feed ingredients are first conditioned with steam and/or water under atmospheric pressure (feed mixture at this stage will contain 20–30 per cent moisture; conditioning temperature 65–95°C) and then conveyed to a pressurised extrusion barrel (known as an extruder) where the feed mixture is then cooked to a temperature of 130–180°C by means of heat and mechanical shear for 10–60 seconds (cooking period and temperature depending on the particle size of the feed ingredients, the composition of the feed mash and the physical property of the extruded diet required). The cooked meal is then extruded via a tapering screw through a die plate at the end of the pressurized extrusion barrel to the exterior where it then expands and is cut into the desired length or physical form. During this process, the extrusion cooked feed emerges from the die with a lower bulk density and having moisture content of 25–30 per cent, which then requires further drying. The extrusion process requires a certain amount of carbohydrate (as starch) to be present; starch on gelatinizing becomes plastic, absorbs water and on superheating vaporizes with consequent expansion.

Advantage of Extrusion-Cooking

The higher temperatures employed during extrusion cooking facilitate the rupture of the cellulose membrane surrounding the plant cell and individual starch granules of cereals and oilseeds, with consequent starch gelatinization and increased carbohydrate and calorific bioavailability.

The higher temperatures employed during extrusion pelleting facilitate the inactivation and/or destruction of the heat-labile anti-nutritional factors usually present within cereals and oilseeds (*i.e.* enzymatic growth inhibitors) and exogenous contaminants within animal by-products.

Extrusion cooking produces feed pellets that are extremely stable in the dry state and thus can be stored for prolonged periods without nutrient degradation.

The higher mechanical durability of extruded pellets (brought about through starch gelatinization and strong intermolecular binding) results in fewer fines being produced during handling, transportation and feeding, and consequently ensures maximum feed intake and minimises water pollution (through the potential decay of uneaten fines within the water body in which the fish or shrimp are cultured.

In contrast to the majority of steam pelleted feeds, extrusion cooked feed pellets are extremely stable in water and will maintain their physical integrity for prolonged periods allowing more feed to be consumed while maintaining water quality; as such extruded feeds are ideally suited for those aquaculture species with slow feeding habits such as marine shrimp.

Extrusion cooking offers the feed manufacturer the flexibity to produce water stable feeds tailored to the physical feeding requirements of the aquaculture species in question (*i.e.* in terms of feed texture, palatability, bouyancy, shape and colour). For example, due to their low bulk density and porous nature, expanded feeds can be rehydrated with 200–300 per cent water (either alone or with dissolved dietary feeding stimulants) prior to feeding so as to produce a soft or moist extruded feed, and/or coated with lipid (either alone or mixed with a vitamin/phospholipid/pigment premix) to produce a high lipid or vitamin protected (through lipid coating or emulsification) diet with goodwater stability and reduced nutrient leaching characteristics. Futhermore, through careful formulation and by controlling starch gelatinization within the extruder barrel, it is possible to produce feeds with different final bulk densities and consequently with either floating or sinking properties. Floating feeds are ideally suited to intensive cage farming activities where feed losses can be kept to a minimum and visual checks made on feed consumption.

However, on the negative side, it should be pointed out that extrusion pelleting is more expensive than regular steam pelleting (both in terms of equipment procurement and operating energy costs, including the added cost of drying the extruded feed) and may result in the loss or damage of heat-sensitive nutrients (*i.e.* such as ascorbic acid, thiamine, poly-unsaturated fatty acids and lysine) if cooking is not correctly controlled rather than over fortifying the feed mash before extrusion cooking, heat sensitive additives (*i.e.* marine lipids, vitamins, antioxidants, emulsifiers, and pigments) may be sprayed onto expanded pellets after extrusion. Further-more, in view of the high carbohydrate requirement (c. 15-25 per cent diet) for adequate extrusion cooking, care should be taken when fixing dietary carbohydrate levels within formulations intended for carnivorous fish or shrimp species, which have a low dietary tolerance for digestible carbohydrate.

Cooler and Dryer

Temperature and moisture from the dry pelleting processes should be reduced within few minutes after extrusion. This is done by collecting the pellets and spreading them in thin layer on a concrete or tiled floor and blowing air over them or by solar drying. In commercial feed production, vertical or horizontal type cooler – dryer with an circulation are used. Temperature is reduced to ambient levels and moisture preferable is below 10 per cent but not exceeding 13 per cent Cooling and drying is done by passing the hot pellet through vertical or a horizontal chamber designed to bring air at ambient temperature into intimate contact with the outer surface of the pellet produced.

Vertical Cooler Dryer

Feed pellets in this dryer are discharged from the mill into the top of a flat sided hopper and dropped into an attached cooling bin. The bin is divided in the middle with a plenum (Sealed chamber containing pressurized air) connected to the suction side of a blower fan. The weight of pellets filling the cooler central pin perforated lowers on the two sides to allow cool air to permeate the hot pellets, removing moisture and cooling the pellets before entering the plenum for discharge through the blower, pellets leave the bin of the bottom via discharge of gates at a rate regulated by the amount of hot pellets entering the cooler. This ensures uniform cooling and drying of pellets.

Horizontal Cooler Dryer

Coolers of the horizontal type consists of a moving wire belt or sectional belt of perforated metal trays which convey pellets from the discharge spout point of the pellet mill. The depth of pellets on this belt and their speed of travel are so adjusted that pellets leave for storage at a desired moisture and temperature. Horizontal coolers may be a single dock with pellets discharged at the end opposite the intake, or a double dock with two belts in the same enclosure, pellets return to the same and as they entered. Air from a centrifugal fan is made to flow from the cooler bottom through the layer of pellets.

Solar Feed Drier

Sun-drying is one way of drying moist feed to form a dry product which can be stored. Two of the problems in sun-drying are that the

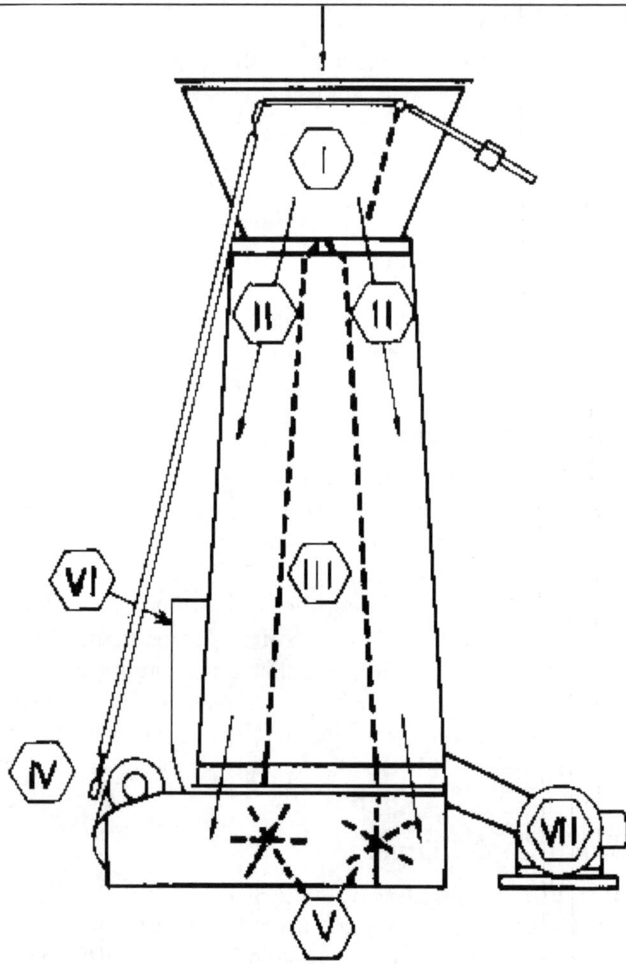

I Hopper with Device to Regulate Pellet Flow
II Cooling Areas
III Air Chamber
IV Discharge Mechanism Motor
V Discharge Gates
VI Centrifugal Fan
VII Fan Motor

Vertical Cooler/Drier

feed attracts flies during drying and that, in seasons when there is intermittent rain, it is difficult to complete the drying satisfactorily even though the sun is very powerful during the rest of the day.

A solar drier developed for drying fish for human consumption (Thomson, 1979) can be adapted for use for drying fish feeds. The principle is shown in Figure. Basically, any greenhouse type unit where convection occurs is suitable. Air should be channelled over shallow layers of feed placed on a dark surface. The scale of the drier can be adapted to the daily feed output. A clear cover which slopes to the sides as well as to one end would be better than the design illustrated in Figure, which is prone to leakage during rain. It is hoped that the illustration will however provoke ideas for adapting the principle involved for drying moist fish feeds for storage.

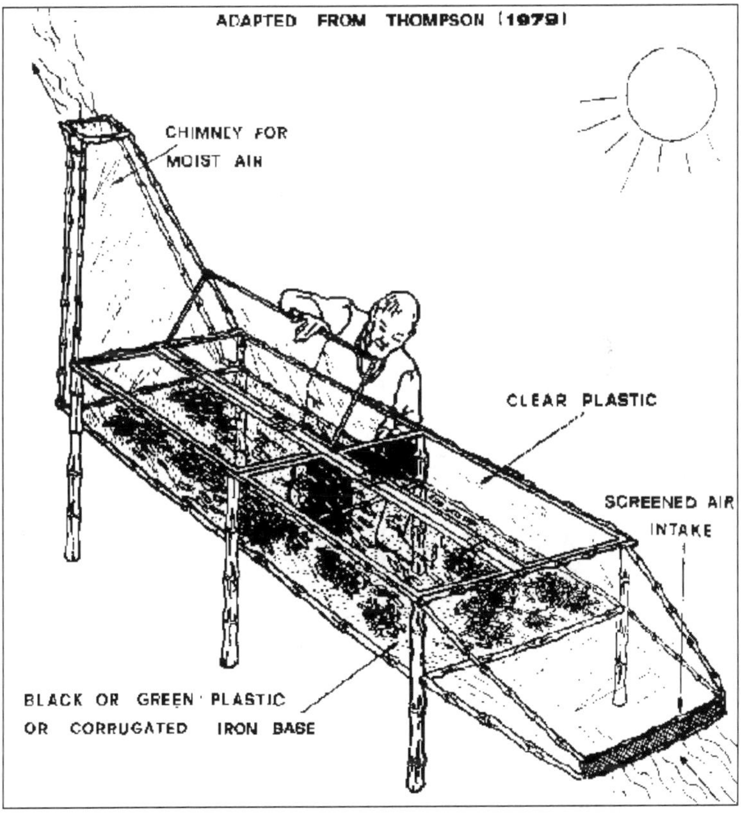

Solar Feed Drier

Surge Bins

This term simply refers to temporary storage bins where feed can be held stationary while awaiting movement to the next stage of the manufacturing process. For example, where a batch mixer takes 15-20 minutes to mix a feed, the presence of a surge bin before it enables the next set of ingredients to be weighed and transferred into it during the mixing process of the previous batch. Surge bins are a means of speeding up production rather than gearing the whole output of the mill to the slowest part of the process. (It is a temporary storage bins where feed in held in stationary while waiting for next stage of feed manufacturing process).

Fat Sprayer

Some times, fat is added to feed after pelleting or extrusion, because high lipid feed do not pellet, so well as low lipid feed. Fat is therefore sprayed into the feed in mixers placed in the production flow after pelleting equipments. After pelleting, high lipid levels are sprayed.

Cookers

Cookers are used for cooking or steaming of feed which increases water stability through starch gelatinisation. Obviously, the design of cookers has almost infinite variation. Hastings (1975) mentions two types capable of steaming up to 10 kg batches of moist feed cakes, made by hand. One is a cylindrical kettle with a false perforated bottom on which the cakes can be placed and steam allowed to enter. The other type has an hour-glass shape and the cakes are placed above a bamboo screen placed over the narrowest section.

Steam Boiler

Where steam is used in the pelleting process, a steam boiler is needed. A pellet mill with a capacity of 1.0-1.5 tons per hour of pellets would require a steam generation plant capable of producing about 60-90 kg/hour of steam at 100-150 psi. It is kneader where steam is used in the pelleting process.

Crumbler

A crumbler is a roller mill with rolls specially designed for breaking up pellets into smaller particles (Figure). Usually the

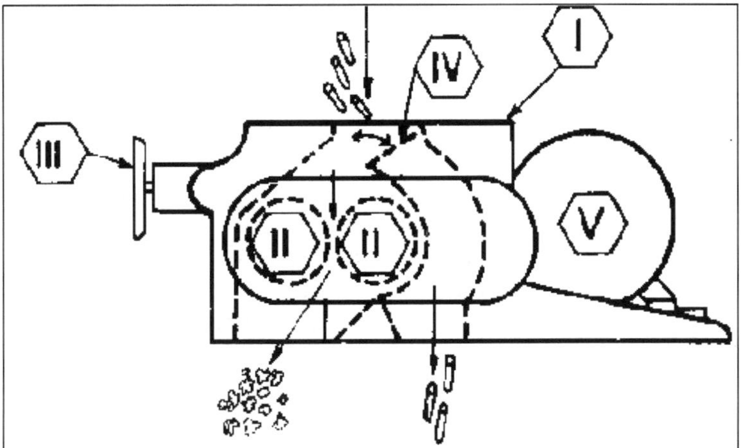

Crumbler

crumbler consists of two corrugated rolls situated below the cooler/ drier exit. The pellets can then be diverted into the crumbler, if crumbles or granules are desired, or they can by-pass it. It is a roller mill with rolls for breaking up pellets into smaller particles. It is used for breaking up pellets into smaller particles. It consists of two corrugated rolls situated below the cooler/drier exist. The pellets can live feed into crumbler, if crumbles or grawles are desired or they can be pass it. It is necessary to produce granules or crumbles for feeding small fish. Cooled pellets are ground by compared rolls (roller mill) and gravel.

Freezer

A freezer will only be required if it is intended to freeze moist feed or moist feed ingredients for subsequent cold storage. Space can often be rented at a local cold store or fish processing unit; it is doubtful if the purchase of a cold store could be economic for a small-scale aquaculture feed plant.

Sifter

The sifter is a separator, usually oscillating, with a number of screens. It is used to separate off crumbles or granules which are too large and, both in pellet and crumble manufacture, to screen off the dusty portion (fines) of the feed for return to the pelleter for further processing. The sifter is a means of ensuring a good quality product with the right particle size and a low level of fines, which are wasteful to feed. It is a separator, usually oscillating with a number of screens. It is used to separate crumbles or granules which are too large and both in pellet and crumble manufacture to screen off dusty partial (fines) if the feed. The sifter is a means of ensuring a good quality product with the right particle size and a low level of fines which are waste feed to feed.

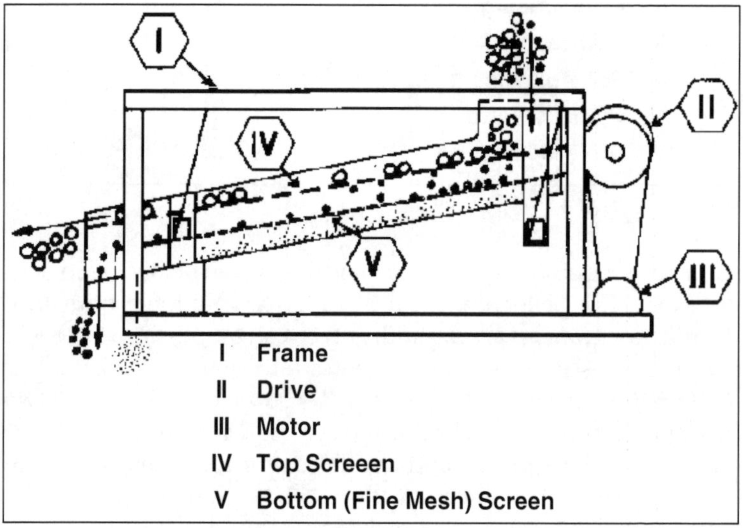

Sifter

Bag sewer

It is used for sewing the top portion of the bags. Machines are available for sewing the tops of the bags in which dry feed may be placed for storage. This machine is not normally necessary unless the feed is going to be sold to other farms.

Scales

Accurate scales for weighing ingredients and completed feeds are essential parts of all feed mills. For the size of operations being considered here, simple platform scales are adequate. The type that have a direct reading dial are the easiest to use and are least likely to cause mistakes to be made. Those which require manual balancing by moving counter weights along a bar are less reliable, but cheaper. Accurate manufacture according to formulation depends on good scales; this item is often neglected, with insufficient money being spent on it. Scales with a taring device are the best (the ability to adjust to zero after a container has been placed on them so that the weight of the actual ingredient being weighed can be read directly from the dial). Accurate weighing for ingredients is essentials. Scales with taring device are the best.

Elevators and Conveyors

There are many different types for conveying feeds from one part of a mill to another or from one piece of equipment to another.

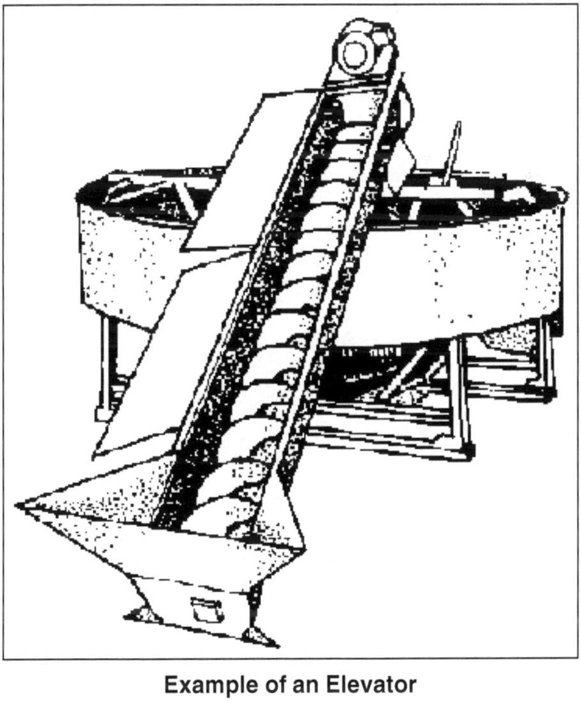

Example of an Elevator

Types of Equipment Necessary for the Production of Different Kinds of Aqua Feeds

	Dry				Non–Dry		
	Mash Meal	*Floating Pellets*	*Sinking Pellets*	*Granules*	*Extruded*	*Pastes, Cakes and Balls*	*Non-Formed*
Grinder/Mill (Dry Products) 1/	+	+	+	+	+	+	–
Grinder (Wet Products)	–	–	–	–	+	+	+
Dry Mixer	+	+	+	+	+	+	–
Wet Mixer	–	–	–	–	+	+	?
Elevators 2/	?	+	?	?	?	–	–
Conveyors 2/	?	+	?	?	?	–	–
Pelleter and Dies	–	–	+	+	–	–	–
Mincer/Extruder and Dies	–	–	–	–	+	? 3/	–
Cooker/Extruders (Expanders) and Dies	–	+	–	–	–	–	–
Surge Bins	?	+	+	+	?	–	–
Cooler/Dryer	–	+	+	+	? 4/	–	–
Fat Sprayer	–	?	?	?	–	–	–
Cooker	–	+	–	–	?	?	?
Steam Boiler	–	+	+ 5/	+ 5/	–	–	–

Contd...

Contd...

	Dry				Non–Dry		
	Mash Meal	Floating Pellets	Sinking Pellets	Granules	Extruded	Pastes, Cakes and Balls	Non-Formed
Freezer	–	–	–	–	? 6/	? 6/	? 6/
Crumbier	–	–	–	+	–	–	–
Sifter	–	+	+	+	–	–	–
Bag Sewer	?	+	?	?	–	–	–
Scales	+	+	+	+	+	+	+

1/: Not needed if all dry products purchased in ground form; 2/: Manual labour can substitute, especially in smaller plants; 3/: Feed balls sometimes formed from extruded products; 4/: Only if product not used moist; 5/: Not in 'cold' process; 6/: Only if product not to be used immediately.

Some are operated horizontally, some on a slope, others elevate the material vertically toe different level of the automated building. The requirement of elevators and conveyors will depend on how automated the mill is to be and how its plant is laid out. It is used to minimize labour needs and maximize the use of land by building multistorage rather than horizontally arranged plants. There are many different types of equipment for conveying feeds from one part of a mill to another or from one piece of equipment to another. Some are designed to operate horizontally, some on a slope (Figure), others elevate the material vertically to a different level of the building. The requirement for elevators and conveyors will depend on how automated the mill is to be and how its plant is laid out. The utmost use of such equipment is made in the modern animal feed mill (often using pneumatic systems) to minimize labour needs and to maximise the use of land by building multi-storey, rather than horizontally arranged plants. In small-scale aquaculture feed mills, most of this type of work (feed transfer) will be done manually so the topic of mechanical equipment for this purpose is not dwelt upon further here (see 'further reading' at the end of this section).

Equipment Sizing

Having decided to set up a plant to produce a certain type of aquaculture feed you then have to decide what size of equipment to purchase. This must be done by working out the maximum amount of feed you need to manufacture in one hour. The following example is provided.

We will assume that you have decided to make a moist feed on a daily basis. Also that you are unable to even out your farm production/harvesting schedule completely because of seasonal differences in growth rate and the fact that you do not have many ponds. Let's say that you have therefore worked out, from typical AFCR's and from feeding rate tables, that you will require between a minimum of 2 tons and a maximum of 8 tons of moist feed per day.

If you are going to produce a dry pellet, your needs could be averaged out (to say 6 tons/day). You could then decide only to work a 7-hour day, 5-day week, 50-week year, building up a stockpile to serve the period when feed demand is greatest. You would then need to size your plant on the basis of

$$\frac{6 \times 365}{7 \times 5 \times 50} = 1.25 \text{ tons/hour}$$

A pelleter capable of producing 1.5 tons/hour would be adequate. Bear in mind that large pellets can be produced more quickly than small pellets so the final sizing of the plant depends on the rate of production of the equipment when it is set to make the sizes of feed you intend to use on your farm.

You have a number of options here. One would be to size your plant so that it is capable of producing 4 tons/day in an 8 hour work shift and, when the demand is greatest, work on a double shift basis. Whether it is best to do that or to size the plant to a capacity of 8 tons/day in an 8-hour shift depends on the local economics of labour and equipment costs. Installing a plant which is capable of producing twice as much feed in a given time does not necessarily double its capital cost.

We will assume, for this example that you have decided to produce 8 tons/day in an 8-hour shift (average 1 ton/hour). The major piece of equipment you need to purchase, which governs the rate at which your whole mill operates, is the mincer/extruder. You will need one capable of extruding at least 1 ton/hour of the size of feed which you intend to use. I mentioned earlier that the same type of equipment is also suitable for mincing/grinding your moist raw materials. If (say) 30 per cent of the feed, on average, is composed of moist ingredients you will need an additional mincing capacity of 0.3 tons/hour. Rather than buy one mincer capable of (say) 1.5 tons/hour to cover all your mincing and extrusion needs it would be better to buy two separate mincers, one to be used normally for raw material preparation and one normally for mixed feed extrusion. This gives you greater flexibility and means that the two different operations can proceed simultaneously, rather than sequentially. It also gives you a 'spare' mincer if one should break down, which would enable you to cover that emergency through longer working hours until a repair could be effected.

If you are producing one ton per hour maximum of extruded feed you will also need to produce one ton per hour of mixed feed to put into the mincer. However, this does not necessarily mean that you need a mixer capable of holding a whole ton in each batch. (Mixing is a batch operation, as compared to extrusion, which is a continuous operation). Again, the final sizing depends on the economics of labour versus equipment costs. It may be better, since at least two batches of feed could be mixed during one hour, to buy a wet mixer capable of mixing 0.5 ton batches. Before the wet mixture is prepared, the dry ingredients may be mixed separately in some processes. As the same kind of mixer will do in both cases, it may be better to buy two small mixers, rather than one large one, from the repair point of view (as in the case of mincers). If the moist ingredients represent 30 per cent of the final product, it follows that the dry ingredients will constitute only 70 per cent of the mixing volume if you wish to premix them. In the example given here, I would suggest the purchase of two mixers, each capable of mixing batches in the 0.5–0.75 ton range. Going back a stage further in the manufacturing process become to dry grinding. You must assess what proportion of your ingredients will be delivered in small enough particle size for immediate use and which will have to be ground on the farm site. Let us assume that 50 per cent will have to be ground or re-ground. You would then need to grind (assuming a year-round average of 6 tons/day of feed $\frac{6 \times 365 \times 70 \times 50}{100 \times 100} = 765.5$ consumption) tons of dry ingredients per year. Again, you have a number of options. You could grind every day, on a 5-day week, 7-hour day, 50-week year basis, in which case you would need a grinder capable of grinding about 0.5 ton/hour of the ingredient which you have in your formula which is the most difficult (*i.e.*, the slowest) to grind. However, grinding is a noisy, dusty and unpopular activity. Also, grinders are relatively cheap so you might decide to buy a grinder capable of grinding 5 tons/hour and only use it two days per month. Against this, you must balance the frequency of raw material delivery and the fact that ground materials do not have such good keeping qualities as unground feeds.

As you will see from the above example, there are no strict rules for the sizing of the equipment for your feed plant; many options are open. However, I hope the example and discussion has given you some ideas on how to approach the topic. Start by calculating your maximum daily feed production requirement and work back from the size of the main piece of equipment (pelleter or mincer for formed diets, or mixer in the case of meals or mashes), balancing the important factors shown below:

- ☆ Ease of operation
- ☆ Relative cost of options
- ☆ Back-up capability in case of equipment failure
- ☆ Storage requirements and availability
- ☆ Maximum use of equipment

Finally, a word or two about cost. Any costs of equipment quoted in this manual would rapidly become out of date, so no specific details are given. Equipment suppliers would be able to supply this data at the time the feed plant is being designed. However, the following examples are provided to give some idea of the scale of cost involved in feed manufacture:

1. The equipment necessary (based on manual labour) for a simple mixing and extrusion plant for producing approximately 5 tons per day of moist feed (excluding the grinding of dry materials) was estimated to cost about US$ 20,000–25,000 (1984 prices). This figure did not include buildings, power supplies etc.
2. A complete 5 tons/day moist feed plant, operating on a 6-8 hour/day basis, inclusive of all grinders, mixers, extruder, conveyors, steelwork, electrical controls etc., but excluding erection costs, building, and commissioning, was quoted at about US$ 160 000 (1984 prices). This plant was based on minimization of labour.
3. A complete 5 tons/day dry pelleting plant as in (b) was estimated at about US$ 140 000 (1984 prices)
4. A simple lay-out consisting of a vertical mixer, a hammer mill, a 1 ton/hour pelleting system, including pelleter, conditioner, cooler, surge bins and crumbler, one bucket elevator, and three screw conveyors, but excluding steam boiler, was estimated at US$ 71 000 (1982 prices) for India.

One more point–don't forget to buy adequate spares for all your machinery. No machine works for ever. Make sure you have spares on site of all components which are most likely to wear or break.

Storage

It is meant for both feed ingredients and finished feeds. It is an important step in maintaining the quality of the feeds. Vitamins and mixture should be kept in coolest place and used within 6 months. Lipids should be stabilized with Antioxidants and kept in sealed dark plastic container in a cool place. Dry ingredients and feeds should be used within 2-3 months.

Chapter 20

Feed Handling and Storage

A manufactured diet requires storage at least at the place of manufacture and on the farm. Feeds are composed by perishable material which deteriorates with storage. Therefore, it is always desirable to minimize storage time.

When feeds are stored for too long periods (or) under poor condition, serious problems arise. Losses occurring in feed stuffs during storage fall under 4 groups. Weight loss, quality loss, health risk, economic loss. These losses arise from cause of deterioration of feed during storage.

Oxidative damage	Microbial damage	Insect/indent damage	Others chemical changes
Effects			Effects
Oxidative rancidity	1. Reduced nutritional value 2. Non enzymatic browning 3. Poor flavour and appearance, making feeds lump and less palatable 4. Mycotoxins (*Aspergillus flavus* cause cancer (GNOC, cotton seed)	Produce heat and increasing moisture (for bacterial invasion)	Chemical produced may reduce amino acid vitamin availability

Losses and Deteriorative Changes which Occur during Feed Storage

Environmental factors, such as moisture (feed moisture content and relative humidity), temperature, light, and oxygen influence deteriorative changes and losses in feedstuffs. These affect the feedstuff either directly or by influencing the rate of development of insects and fungi, which consume the feed during storage.

The following are the major factors which affect the quality and weight of feedstuffs during storage:

1. Major losses due to human theft, fire and the consumption of scavenging animals such as rats and birds
2. Damage due to rain and condensation and to high temperatures
3. Damage by insects and damage by fungi,
4. Changes in the quality of the feeds due to enzymatic actions and the development of oxidative rancidity

Of the above factors, damage due to rain and condensation and to high temperatures is probably the most important as it influences the rate of loss or damage caused by most of the others listed. Though oxygen (from air) is necessary for the development of oxidative rancidity and for the growth of fungi and insects, it is impracticable to exclude it from feed storage areas. Oxygen is sometimes replaced by nitrogen or carbon dioxide for the storage of specialized foods for humans, but has to be accepted as omnipresent in feed stores.

Physical Loss

Significant loss can occur as an accumulative effect of individually small, but regular theft. Less obvious are the losses caused by scavenging animals, particularly rats and mice. Food stores are notorious breeding grounds for such animals.

Temperature increases sufficient to cause fire can occur in stacked feeds. Feed stores are flammable, particularly if they are constantly full of fine atmospheric dust from grinding processes within the store or adjacent areas.

Water and Heat Damage

High levels of moisture content and relative humidity cause direct losses by making it difficult to use the material in its original form. It may be too wet to mix if it is an ingredient, or its physical structure may be destroyed if, for example, it is in pelleted form. More serious is the effect that high levels of product moisture and relative humidity have on insect infestation and the growth of fungi.

Regardless of the initial moisture content of the feed materials put into the store, their actual moisture content will gradually reach an equilibrium dependent on the relative humidity of the air in the store. Generally, a safe moisture level for a specific product is that which develops at a relative humidity of 75 per cent However, relative humidities above this level in tropical areas are common; moisture levels in feeds therefore tend to rise. This is one of the reasons why it is sensible to store feeds for much shorter periods before use in tropical areas than in temperate zones. Cereals will store quite well at 10-12 per cent moisture. In general, moisture levels of 10 per cent or less should be sought. Fungal growth increases moisture content also.

High temperatures also affect the rate of loss and damage in feeds, another reason why feeds in tropical zones should not be stored as long as in temperate areas. High temperatures in feeds may occur not only because of environment and the way in which they are stored but because of the heat generated by the growth of fungi and insects. Increases in temperature within large stacks of feed have been known to cause 'spontaneous combustion' followed by fire losses. Increase in temperature may reduce the availability of the amino acids in feeds.

Insect Damage

Feeds are attractive places for insects, including various species of moths, weevils and beetles, which consume the feed. All grow well at normal temperatures in feed stores. At temperatures from 26-37°C, they can reach epidemic proportions. Insects thrive better on ground materials. Whole cereals or oil cakes can therefore be stored longer than meals made from them. Insects cause damage through weight loss, the exposure of the feed ιo further damage by fungi and through oxidation and the introduction of contaminating bacteria.

Fungal Damage

Fungi grow at relative humidities above 65 per cent, moisture contents generally above 15 per cent (although some mycotoxin producing fungi grow well at only 9-10 per cent moisture) and temperatures which are specific to the fungal species. Most fungal growth occurs at temperatures above 25° C and relative humidities above 85 per cent. Higher temperatures and moisture levels favour increased growth. Fungal growth itself encourages local rises in temperature and moisture content. Many fungi are killed during the processing of ingredients but their spores are resistant and remain present to re-infect the material later if the environmental conditions become favourable for their development.

Fungal growth causes weight loss, increases in temperature and moisture, staleness (off-flavour), discolouration and, perhaps worst of all, some common species produce mycotoxins. Mycotoxins, the best known of which are called aflatoxins, are known to be toxic to some species of fish at least. In addition, as the toxins remain in the flesh of the animal which consumes them, they are also a health hazard to humans. Sorghum, maize and its by-products, groundnut, cottonseed, cassava, coconut and sunflower are ingredients especially prone to contamination with mycotoxins.

Chemical Changes during Storage

The following deteriorative changes can occur in feeds during storage.

Lipids can break down into free fatty acids which make the feeds more prone to the development of rancidity. This breakdown can be caused by the damage resulting from insect infestation and fungal growth. High lipid ingredients are more susceptible to this type of chemical change than others. Carbohydrates may ferment, to produce alcohols and volatile fatty acids.

Lipids undergo oxidation, causing rancidity. Materials with high levels of poly-unsaturated fatty acids and of course, pure lipids themselves, are more prone to the development of rancidity than others. Fish meals, expeller vegetable oil cakes and rice bran are particularly vulnerable. Grains have natural antioxidants which protect them from rapid deterioration. There are various types of chemical reactions which cause rancidity to develop. The result, as far as feed quality is concerned, is similar. Rancid fats reduce the palatability of the feed and contain toxic chemicals which may depress growth. Chemicals may also be produced which reduce the availability of amino acids in the feed proteins.

Vitamin potency decreases during storage (and processing), particularly in premixes which also contain minerals. Naturally occurring vitamins in feed-stuffs also deteriorate on storage. Vitamin C is particularly susceptible, as is thiamine (vitamin B_1).

Good storage should give protection against high temperature, humidity, moisture insect and rodents; fresh fish should be used immediately or kept in frozen until use.

Feed Storage Facilities

1. Insulation
2. Ventilation
3. Storage hoppers and silos.

Insulation

Insulation is done to prevent extreme heat or cold. Choice of materials for insulation should be with poor particle contact characters and poor condition. Insulation materials should be waterproof and vapour proof.

$$Q - UA\ (t_1 - t_0)$$

where,

- Q: Heat loss or gain (Qty of heat passing)
- U: Thermal transmittance (or) Heat passage
- A: Structure Area (m^2)
- T_1: Temperature inside
- T_0: Temperature outside
- U: Heat passage through a surface area under pressure of given temperature gradient. It is represented as watts/°C m^2. Higher the value of U, the performance of insulation material is poor.

The value of U for certain construction building materials are given below:

Wall Construction Material	*U Value*
215mm dense hollow concrete block work	2.05
215mm solid wall of framed block work + rendering both sides	0.45
6mm cement fibre sandwich on 50 x 50mm timber frame with expanded polyester core	0.45

Wall Construction Material	U Value
Roof	
Corrugated cement fiber sheet	6.53
Corrugated sheet steel + fiber insulation board	4.52
Corrugated cement fiber sheet + 40mm foil faced Polyurethane board	0.5
Floor	R (°C m²/W)
Dense Concrete floor	0.042
Concrete slatted panel	0.086
150mm light weight aggregate	0.17
Wooden slate 58 × 70mm with 10mm gaps	0.23

Thus, the construction material should have low 'U' value and high R value. Usually in insulated stores, IR emitters are used to avoid freezing temperature and AC units are used to avoid high temperature.

- ☆ High U – Poorer performance as Insulators
- ☆ Low R – Poorer performance as Insulators
- ☆ Low U and High R – always better

Ventilation

Ventilation or air exchange should be provided to lower temperature and to control humidity.

Ventilation

- Natural means (open front/ open ridge)
  - Most commonly used
- Partly controlled (exhaust fan)
  - Have more control But, not suitable for partially open building with adverse temperature gradients.
- Totally controlled (using complex duct/Pressurizing fan equipments)
  - Most costly and carefully designed to distribute the air evenly

Heat exchanged through ventilation may be shown as:

Heat loss or gain (Q) = $m \times s \times (t_e - t_1)$

where,

m: Mass of air exchanged/unit time

s: Specific heat of air (0.36µ/kg)

t_e: Temperature of exhaust

t_1: Temperature of inlet

Storage Hoppers and Silos

1. Bulk storage systems are widely used for storing dry aqua feeds, silages and binder meals. Silos used for storing dry feeds ranges and are typically prefabricated in galvanized steel in volume from 2-30m^3. Silos used for storing silages are generally larger, upto 60m^3
2. Silos are internally coated with acid resistant material in order to prevent corrosion.
3. Silos are available with large diameter catches to accommodate filling from bulk sacks.
4. Problems associated with dust and crumbs in bulk feeds can be reduced by the use of bucket type of evaluators. These are pore expensive than the normal pneumatic filling system. But, reduce physical damage to pellets. The disadvantage in the use of bulk storage which reduce the flexibility for storing of feeds.

General Recommendations for Dry Storage–"Do's" and "Don'ts"

For simplicity, this section of the manual has been arranged in the form of a series of instructions:

PROVIDE a building for storage which is secure and can be adequately locked. Ensure that its roof will protect the feed from rain and that surface water cannot enter the store. Provide it with ventilation points (windows are not necessary or recommended). Ventilation entry points should be low on the side facing the prevailing wind and high on the opposite side. Orient the building so that one of the long sides faces the prevailing wind. Ensure that all entry points are meshed to prevent entry by birds, rats etc. The drier and cooler you can keep this store the better your feed quality will be.

DO NOT accept deliveries of raw materials which are visibly damp or mouldy or which are obviously infested with insects.

PLAN your ingredient purchases carefully so that you do not need to keep too great a quantity in stock. Obviously you will want to store greater quantities of seasonally cheap or scarce materials. But, do not be tempted to buy a year's supply just because they are cheap now. It may prove very expensive indeed if half of them have to be thrown away.

ALWAYS keep the store clean. Floors and walls should be regularly swept. Spilled material must be removed and the contents of broken bags or containers used first. Cleared areas of the store must always be cleaned before new materials are placed there.

ARRANGE your store so that new deliveries are not put in front of old stocks. The oldest materials MUST be used first.

MAKE small stacks. Large stacks of sacks lessen insect damage, which occurs mainly at the surface, but cause heat generation, with other consequential damage. In the tropics, it believed that small stacks which are used rapidly are better than large ones which remain stagnant for long periods. If possible, RAISE the sacks off the ground by stacking them on wooden pallets (platforms).

ENSURE that ingredients are clearly and indelibly labelled so that those drawing from the store are sure that they are drawing the correct ingredient (some look very similar when ground) from the oldest batch.

DON'T walk on the stacks of compounded feeds unnecessarily. This will break the pellets on the surface and lead to the production of a lot of wasteful fines (dust).

DON'T allow sacks to rest against the outer walls of the store–leave a space between the stacks and the wall.

DON'T allow staff to sleep or eat in the feed store nor, preferably, to smoke.

Feed sacks do not touch the floor or sidewalls; Stored on the pallets, off the ground and away from wall to allow air to circulate and maintain even temperatures.

Should not be stored in direct sun light. (This would adversely affect the vitamins and lipid quality of the feed) and for long periods. Feeds should be used within 2-3 months. Feed stores should be 100 per cent water proof and damp proof. Sacks arranged properly for easy access. Large stacks reduce or minimize insect damage, but have negative effect by which more heat is generated. Proper ventilation should be provided. Good storage should provide protection against high temperature, humidity, moisture and insect and rodents infestations. Dry feed should be stored under cool, dry conditions, ideally at temperature below 20°C the and RH below 70 per cent

Storage Procedures

Many problems can occur during feed storage. Some deterioration is inevitable. Thus, ingredients should be stored for as short a period as possible and compounded feeds used quickly, especially in tropical conditions. The method of storage depends on the type of ingredient.

Vitamins

Vitamins and vitamin mixes are extremely expensive ingredients and should be given special care. Their volume is usually small because their inclusion rate is low, so storage space is not normally a limiting factor. Vitamins should not be mixed with minerals before storage. Vitamins and vitamin premixes should be kept either in the manufacturer's containers or in air-tight light-proof containers. They should not be kept in hot sunny rooms. They should be kept in the coolest place available, preferably under air-conditioning. Stocks should be turned over at least every six months.

Moist and Wet Ingredients and Compounds

These materials, such as 'trash' fish, must either be used fresh (they can be iced down for up to 12 hours while stocks are being drawn from a delivery for use) or they must be kept frozen. Proper freezing of large quantities of wet materials like these requires a plate freezer and, unless you already have that kind of facility (say for the freezing of your harvested aquaculture product), freezing should be contracted out to a specialist. Once frozen, this type of ingredient can be stored in a cold store. The temperature of the cold store should not be higher than –20°C, preferably below –30°C. Rules must be established to minimize the time that the cold store door remains open and the number of times it is opened, or the product temperature will rise too high. If access to plate freezing facilities run by specialists is unavailable, small quantities of wet materials can be frozen in a cold store. Only very small quantities, though. The material must be spread into thin (1 inch) layers and the store must not be overloaded with huge quantities of unfrozen materials; it will be unable to cope. If you simply put a large drum or bag of trash fish in the cold store, it will be several days before it completely freezes; rapid deteriorative changes in the feed will be taking place all that time. Moist compound feeds should be used on the day of manufacture within 2-3 hours. Don't use wet ingredients which smell of ammonia.

Dry Ingredients (Raw Materials)

Dry Compounded Feeds

These should be treated in the same way as dry ingredients but should not be stored for such a long time. Mixed feeds are more prone to damage than individual ingredients. This is because of interactions between different ingredients and because of cross contamination with insects and fungi. Mixed feeds which have undergone a heat treatment during production, such as steam pelleting, store better than other mixtures because many of the damaging factors will have been destroyed.

Storage Time for Selected Feedstuffs

Types of Feed/Feeding	*Time of Storage in Tropical Counters (months)*	*Time of Storage in Temperate Counters (months)*
Ground ingredient	1-2	3
Whole gain and oil cakes	3-4	5-6
Compounded dry feeds	1-2	1-2
Vitamin mix (kept cool)	6	6
Wet ingredient	2-3	2-3
Frozen material	2-3	2-3

Lipids

Keep in sealed, preferably plastic, containers, in a cool dark place. Ensure that they have had antioxidants added to them when manufactured.

Molasses

No special care necessary in tropical countries but, in temperate zones, molasses may require heating in winter before this product can be used in a mixed feed.

Self Assessment Questions

1. What are losses occurring in feed stuffs during storage?
2. Write detailed account on feed storage facilities.
3. "Do's" and "Don'ts" in general recommendations for Dry Storage.
4. How to store dry and wet ingredients?
5. Storage time for selected feedstuffs.
6. Collect used feed bags from a farm and study the storage condition.

Chapter 21

Anti-Nutritional Factors (or) Antimetabolites and Toxic Factors

It is constituent responsible for deleterious effects and are called as Anti-Nutritional Factors. It is also called antimetabolites. Anti-Nutritional Factors (ANF's) are those one generated from natural plant feeds stuffs exerting inimical *i.e.* adverse effect. It is also a naturally occurring substances in plant materials and greatly influences a metabolic path way. Most of the ANF's are heat labile. The effects are inhibition of growth, decrease in food conversion efficiency, pancreatic hypertrophy, Liver damage and other pathological damages. The seeds of legumes and other plants contain a wide variety of toxic and potentially toxic substances

Classification of Anti-Nutritional Factors (or) Antimetabolites and Toxic factors

Based on nutrients affected and biological response produced in fish, the toxic factors classified as follows:

1. Substances depressing digestion (or) metabolic utilization are Protease inhibitors (PI's), Lectins (Haemagglutinins), Saponins, Polyphenolic compounds (Tannins),
2. Substances reducing the solubility of nutrients are Phytic acids, Oxalic acids, Glucosinolates, Gossypols

3. Substances which inactivating vitamins and hormones (Antivitamin A D E K and Phyridoxine)
4. Mimosine (anti – hormone)
5. Cyanogens
6. Moulds and toxins
7. Bacterial infection (Salmonella contamination and mycobacteriosis)
8. Bacterial toxins (Botulism)
9. Chemical contaminants – pesticides, herbicides, (PCBs, Organochlorines), Heavy metals (Hg), Volatile N – nitrosamines (VNA).
10. Peroxides

Protease Inhibitors (PI's)

Most well known is Trypsin Inhibitors. The most commonly encountered class of feed toxins of plant origin is the protease inhibitors. Most well known among these is the trypsin inhibitor occurring in raw legume seeds, particularly soybean. The mode of action of trypsin inhibitor has not been studied in fish. In other laboratory animals, it is believed to lead to pancreatic hypertrophy with concomitant loss of endogenous protein secreted by the pancreas. This protein is largely made up of cystine-rich pancreatic enzymes. Trypsin inhibitors were first described in soybean, but are also known to be present in groundnut and most other leguminous seeds. Trypsin inhibitors are heat labile. The degree of deactivation of the toxin depends upon the applied temperature, duration of heating, particle size and moisture conditions. Soybean trypsin inhibitors are totally deactivated by moist heat at 100°C for 15-20 minutes. Overnight soaking or germination of the beans, while bringing about improvement in their nutritive value, does not destroy the trypsin inhibitors.

Lectins (Haemagglutinins)

These toxins, which cause red blood cells to agglutinate, are present in almost all legume seeds. The potency of the toxin, however, varies from species to species, with those of *Phaseolus vulgaris* among the most potent. Phyto-haemagglutinins are also heat labile, but care should be taken to ensure that adequate heat (normal boiling temperature) is applied.

These are proteins that posses a specific affinity for sugar molecules. As the name implies, they caurse agglutination of RBC. It is found in seeds of higher plants, Tubers plant saps (rice, wheat, potato, ground nut oil cake, kidney bean). Lectins reduces absorption of nutrients in the gut and cause internal hemorrhages and growth. It is heat labile; cooking or autoclaving is the method used to remove this substance.

Saponins

Three significant characteristics is exhibited by saponins. They are, bitter taste; forming in aqueous solution; haemolysis of RBC. Effects – excess salivation; increased respiration – haemolysis; damage to liver and kidney. Examples are soybean, mung bean, jack bean, alfalfa. Soaking in water for 24 hrs will eliminate this substance.

Tannins

It is a polyphenolic compounds and same effect as saponins. soaking and cooking is the method to remove this substance. It is found in sorghum, mustard seed and rape seed.

Phytic Acid (Phytates)

Plants which are rich in proteins and also rich in phytates. Phytates are salts of phytic acid. It occurs in plants. Phytates form complexes with transitional mineral elements (Zn, Fe, Mn) and make insoluble in Intestinal tract. Addition of phytase enzyme externally in the feed may counter act the toxic effect. Eg. Soybean, sesame, rape seed meal

Oxalic Acid (Oxalates)

Both in plants and animals, oxalic acid (oxalates) found in free and salt forms. Eg. Beet, spinach etc. Water soaking in 24 hrs will reduce the oxalic acid content.

Glucosinolates

It is responsible for pungent flavour found in family cruciferae *i.e. Brassica* sp, which include mustard seed, rape seed and cabbage. It depresses the thyroid hormone. Soaking in water or cooking will reduce this content.

Gossypols

It is a polyphenolic pigments in *Gossypium* (cotton plant) and from Malvacea; It reduces amino acid available. There is reduction in succinic dehydrogenase and cytochrome activity. Gossypol reacts with Fe to form inactive ferrous gossypolate complex. Soaking/ cooking will reduce the content.

Antivitamin Factors

Antivitamin D and antivitamin E factors are known to be present in raw soybeans and raw kidney beans respectively. These anti-nutritive factors are destroyed by heat.

Antivitamin B1

In fresh fish, mussels, herrings, clams, shrimp contain thiaminase (B1). In raw soybean an enzyme called lypoxygenase which catalyze the oxidation of carotenoids (Precursor of vitamin A). Autoclaving will reduce the content.

Antivitamin D

Antivitamin D is known to be present in raw soybeans. Isolated soya proteins has rachitogenic activity. Increase of vitamin D level by 8 – 10 folds in the diet will partially eliminate and it is destroyed by heat.

Antivitamin E

Antivitamin E factor is known to be present in raw kidney bean (*Phaseolus vulgaris*). Autoclaving will reduce the content.

Antivitamin K

Sweet clover disease is characterized by fatal hemorrhagic in cattle. It will reduce the prothrombin level and interfering with blood clotting mechanisms. Autoclaving will reduce this content.

Antipyroxidine

Linamarine (antagonist of pyridoxine) has been identified in Lin seed meal. Extract the Lin seed meal with water and autoclaving will reduce this content.

Mimosine (Free Amino Acids)

It is present in Ipil– iphil (or) suba bull and affect the production of thyroxin; Often, mimosine resembles with tyrosine and function as antagonist against to this amino acid. Soaking in water for 24 hrs will reduce this content.

Cyanogens

Cassava and sorghum are the two feeds most often associated with cyanide poisoning in livestock. Other feeds are linseed meal and certain legume seeds, including the lima bean, the kidney bean, the Bengal gram and the red gram. Cooking in water not only destroys the enzymes responsible for cyanide release, but also volatilizes the HCN released

In plants, present in trace amount in the form of cyanogenic glucosides and present in cassava, sorghum, root crops, and kernels. Specific enzyme glucosidase which hydrolyze cyanogenic glucosides to HCN.

Fungal Toxins

Aflatoxins: Produced by the mould *Aspergillus* spp., which grows, particularly in high temperature/high humidity conditions, on ingredients and compound feeds, particularly groundnut meals and cereals.

T-2 toxins: As above, originating from the mould, *Fusarium* spp.

Vomitoxin: Caused by *Fusarium* spp.

General deterioration of feed quality: Caused by *Penicillium* spp. moulds

Bacterial Infection

Salmonella contamination: Contaminated ingredients, especially animal proteins, together with insects and rodent droppings.

Mycobacteriosis: Unpasteurized trash fish and viscera

Bacterial Toxins

Botulism: Trash fish stored anaerobically

Chemical Contaminants

Pesticides and Herbicides, such as organo chlorine and poly chlorinated biphenyls (PCB's). Plant ingredients, contaminated by spraying and accumulation in fish and fish products used as ingredients

Heavy metals,*e.g.*, Mercury Concentrated by animals and plants used as ingredient sources.

Various: contamination of feeds by traces of chemicals used during feed processing *e.g.*, lubricants, fumigants, water softening chemicals, etc.

Volatile N-nitrosamines (VNA)

Fish meal, especially that dried by hot air exhausted from an oil burner (direct heating/drying method)

Peroxides

Oxidized oils: Poorly stored and inadequately protected (by antioxidants) ingredients, particularly those with high poly-unsaturated fatty acid levels such as fish oils and meals.

The cost of production/kg of fish has to be reduced by including unconventional and cheaper feedstuffs. Various physical, chemical and biological, treatments may be tried to eliminate ANF's from unconventional feeds and other plant feed stuffs. External addition enzymes vitamins and minerals may also be tried to combat depressed activity of ANF's. More and more unconventional feedstuffs may be searched out for the presence of ANFs and evolving various detoxification methods. Cheaper methods of detoxification should be found out which should be practically feasible.

Though most of plant feedstuffs contain ANF's, soybean has been added as commercial, with proper pretreatment of unconventional feedstuffs can be used to replace 20 – 30 per cent of fish meal protein without compromising growth.

Self Assessment Questions

1. What is ANFs?
2. Difference between antimetabolites and ANFs.
3. Common detoxification methods available to remove antimetabolite substance.
4. Classify the different ANF's available in plant feedstuffs?
5. Identify the different unconventional feedstuffs available.

Chapter 22

Nutritional Diseases in Fishes

Nutritional disease has been defined as those which can be attributed to deficiency, excess or improper balance of components present in a fish diet. Such diseases usually have a gradual onset because symptoms do not appear until one or more of the components of a diet drop below the critical level of the body reserves. Also, if food contain all necessary components in proper balance, a nutritional disease is possible.

Types of Nutritional Diseases

It falls under four categories, namely:

1. Those arising from **Under nutrition** or dietary deficiencies or imbalances in the major components of food (Macro nutrients – protein, CHO, fat; Micro nutrients – Thyroid tumour, Nutritional gill disease, Vitamin A imbalance, Nutritional anemia).
2. Those arising from **Over nutrition** (amino acid toxicity, oxidation of unsaturated fatty acids, fatty acid toxicity, dietary vitamin toxicity and mineral toxicity),
3. Those arising from **toxic effect of the diet** (mycotoxins, toxic algae, anti-nutritional factors effect, Senecio alkaloids, anthropogenic chemical disease) and

4. **Others Category** (Photosensitizers, sekoke, spleen and liver, SCP lesions, antibiotic and chemotherapeutic disease, binder disease, blue disease).

Those Arising from Under Nutrition

Dietary deficiency diseases are of two types:

1. Deficiency or imbalance of the macronutrients of the diet – protein, CHO, lipid.
2. Deficiency of the micronutrients of the diet – the vitamins and minerals.

Macronutrients – Protein

Reduced growth and body deformities (general symptoms). Essential amino acids deficiency is impaired the growth. It includes dorsal fin erosion with lysine deficiency. Spiral abnormalities associated with tryptophan, leucine, lysine arginine or histidine. Tryptophan deficiency results in scoliosis.

Important Amino Acid Deficiency Symptoms of Fish and Shellfish

Amino Acid	*Deficiency Symptoms*
Methlonine	Cataract
Arginine	Vertebral deformity
Threonine	- do -
Tryptophan	Vertebral deformity Fin erosion, Kidney stones
Valine	- do
Isoleucine	Vertebral deformity Increased mortality
Leucine	- do -
Phenylalanine	- do -
Histidine	Vertebral deformity
Lysine	Dorsal fin erosion, increased mortality

Macronutrients–CHO

Depress the digestion, enlarged livers (general symptoms). Hepatic degenerative changes and glycogen deposition. Sikoki disease is common in carp.

Macronutrients–Fat

Lipoid Hepatic Degenerative Disease

Not uncommon in fishes. First, Liver taken up a yellow brown coloration. Secondly, anemia follows with gill looking pallor. This is mainly because of over feeding.

Micronutrients

1. **Thyroid tumor** – caused due to iodine deficiency; sign and symptoms: abnormal thyroid follicles; species affected: salmonids; remedial: addition of iodine to fish food, water will prevent the occurrence or cause tumor regression.
2. **Nutritional gill disease** –caused due to pantothenic acid deficiency; sign and symptoms: respiratory difficulties in all gills; species affected: salmonids; remedial: addition of pantothenic acid to fish food, water will prevent the occurrence.
3. **Vitamin A imbalance** produces varying degree of syndromes. Its deficiency causes under growth, blindness, hemorrhage at fin base, keratomalacia.
4. **Nutritional Anemia:** Nutritional anemia is caused by folic acid deficiency and has been reported in channel catfish. Anemia is characterized mainly by depressed values of erythrocytes and haemoglobin. These values vary greatly in normal fishes depending on species, age, water temperature and sex.

Nutritional anemia may have two basic causes. One may be the lack of elements permitting fish to produce blood and second, nutritional. Some of the essential vitamins, minerals and possibly also amino acids belong to the first group. Fatty degeneration of the liver. Visceral granuloma, hepatoma in terminal stages and possibly others belong to the second group.

Major Deficiency Symptoms

Vitamin and Mineral of fish and shellfish

Vitamin/Mineral	*Symptoms*
B_{12}	Anemia, fragemented RBC
Biotin	Gill lamellar degeneration blue slime
Folic acid	Anorexia, anaemia
Thiamine	Hyper irritability, edema
Pyridoxine	Spiral swimming gasping
Riboflavin	Cataract, gills anorexia
Pentothenic acid	Clubbed gills, anorexia
Nicotinic acid	Anaemia, skin lesion
Ascorbic acid	Exophthalmia, Scoliosis
Inositol	Anaemia, Anorexia
Choline	Growth reduction
Vitamin A	Cornealopacity. Exophthalmia,
Vitamin D	Liver and muscle fat, Anorexia
Vitamin E	Clubbed gills, fragile RBC
Vitamin K	Increased clotting time, haemorrhagic fin, bases, gills and eyes
Vitamin C	Scoliosis (broken black disesases)
Calcium	Reduced bone ash, skeletal deformity
Phosphorus	Reduced growth, skeletal deformity
Copper	Reduced haemocyte active and resistance to diseases reduced.
Iron	Reduced haematocrit, anaemia
Zinc	Skin and fin erosion skeletal abnormally cataract
Cobalt	Reduced growth fragmented RBC

Over Nutrition/Disorders Related to Dietary Excesses

Not only the deficiencies (under nutrition), but the excesses (over) directly and indirectly enhance the cost of production while not resulting in commensuration growth enhancement.

Amino Acid Toxicity

Amino acid excesses also have been known to produce toxic

effects. Dietary excesses of leucine at 13.4 per cent of the diet of rainbow trout resulted in vertebral deformities and scale loss.

Nutritional pathological conditions due to the consumption of toxic amino acids or their derivatives have also been reported. Reduced growth, decreased food efficiency and even deaths have been recorded due to the toxic, non-protein amino acid, mimosine (present in leguminous plant seeds) and *L. carnavanine* (present in a plant, *Canavalia ensiformis*).

Oxidation of Unsaturated Fatty Acids

Oxidation of unsaturated fatty acids of diets is disorders due to the presence of auto-oxidation products such as peroxides, hydroperoxides, aldehydes and ketones. These free radicals react with other dietary ingredients such as vitamins, proteins and lipids resulting in disorders associated with vitamin, protein (or) lipid deficiencies. Oxidative rancidity of oils in feedstuff may manifest into pathological changes in aquatic animals feed on such diets. These signs include marked congestion, haemorrhages on skin and fin bases, exopthalmia, loss of appetite, depigmentation or change of body colour, fatty livers, gill clubbing, anaemia, muscle damage etc. Oxidation of fats can be prevented by using antioxidants such as BHT, BHA and Ethoxyquin.

Fatty Acid Toxicity

Due to cyclopropenoic acid (cotton seed oils – fatty acids) has been shown to reduce growth and act as a synergist to carcinogenic effects of aflatoxins (toxins produced by fungus growing on stored food or in food ingredients).

Dietary Vitamin Toxicity

Dietary vitamin toxicity of fat soluble vitamins is contrastingly more pronounced in fish than that of water soluble vitamins. Accumulation of vitamins/fat in excess of requirements, leads to hyper-vitaminosis. Incidence of hyper-vitaminosis in field conditions is unlikely. However, hyper vitaminosis has been studied by experimental induction in fish. Possible signs of hyper-vitaminosis are reduced growths, drop in haematocrit values, fin necrosis, dark colorations, lethargy, reduced RBC count etc.

Vitamin A excess in the diet cause squamoids metaplasia; spelenomegaly; heptomegaly.

Mineral Toxicity

Due to dietary excesses have also been noticed in intensive culture conditions, possible cause being the use of unconventional sources of feed ingredients and use of fishmeal obtained from fish exposed to metal pollution. Poultry waste of arsenic; pulp waste for leads; fish meal for mercury, cadmium etc. have been the known sources of excesses. Dietary excesses of minerals produce signs of reduced growth, low feed efficiency, kidney stones, (Nephrocalcinosis), vertebral deforming, hyper activity, black tail, anaernia, caudal degeneration alongwith growth retardation and reduced feed efficiency.

Average amount of nutrition should present in the body. The excess amount or decrease amount of nutrient will cause disease in fish.

Those arising from toxic effect of the diet (mycotoxins-red disease), toxic algae, anti-nutritional factors effect, Senecio alkaloids, anthropogenic chemical disease, blue disease, black gill disease)

- ☆ **Black gill disease** – causative agents: chemical contaminants such as Cd, Cu, Zn, K, Ascorbic acid deficiency, high organic load; species affected: *Penaeus monodon;* Gross signs: gill shows reddish brownish to black discoloration, loss of appetite, mortality; preventive measure: avoid over feeding
- ☆ **Blue disease** – causative agents: low levels of carotenoids (astaxanthin); species affected: *Penaeus monodon;* Gross signs: sky blue colour, lethargic, soft shell may occur; preventive measure: addition of carotenoids, provide high quality feed.
- ☆ **Red disease**–causative agents: presence of aflatoxins in feeds, rancid feeds; species affected: *Penaeus monodon;* Gross signs: sudden drop in feed consumption, body weakness, poor growth, reddish colour of faecal matter; preventive measure: use recent manufactured feeds, reduce the organic matter. Occurrence of hepato carcinoma in trout. Aflatoxicosis in Rainbow trout. Renal tubular carcinoma, lymphoma, visceral granuloma.

Visceral Granuloma (mycosis)

This disease was described as a mycosis like granuloma by

Wood *et al.* (1955) and carried out extensive work aiming at isolation of the hypothetical fungus causing this disease. This diseased fish grow progressively weaker, become very anemic and seldom survive the second year. This disease usually begins as barely visible papillae projecting from the serosa of the stomach. If gradually spreads to other internal organs, except liver and lower gut.

Toxic Algae Disease

Toxic algae–Microcystis, Anabaena–cause red tide. acute necrotizing bronchial lesions. severe enteric and hepatic necrosis.

Anti-nutritional Factors Effect Disease

Cottonseeds – 2 toxic components

1. Gossypol causes sudden anorexia and deposition of sudanophilic globules within liver and kidney.
2. Cyclopropenoid fatty acids within the kernel. Powerful synergists of aflatoxin B_1 and its metabolites.

Leucaena Toxins

The seed pods of *Leucaena* or Ipil – ipil, a leguminous shrub – Mimosine.

Poor growth – Inappentence – cachexic condition.

Senecio Alkaloids Disease

It causes megalocytosis of hepatocytes.

Anthropogenic Chemical Disease

Residues of substances such as organochlorines, or organophosphates – occasionally/accidentally over dosing or dietary contamination can result in acute toxic episodes. In such condition, deaths out are usually rapid.

Others (Photosensitizers, Sekoke, Spleen and Liver, SCP Lesions, Antibiotic and Chemotherapeutic Disease, Binders)

Photosensitizers

In the presence of an irradiation, it induces a localized tissue necrosis and cause severe necrosis and ulceration. One such compound is phenothiazine It is used for control of Intestinal parasite *octomitus*.

Sekoke Disease

Incorporation of significant levels of silkworm pupae in the diet, leads to this disease. Bilateral cataract; degenerative changes in extrinsic eye, muscle, retina and choroids are several symptoms.

Spleen and Liver Disease

Feeding of minced animal viscera as a major component of the diet for salmonid culture practice. It is regularly associated with the feeding of pig spleen or horse liver.

This contract was generally higher cellular and proliferative.

SCP Lesions

Feeding at high levels of SCP leads to neoplastic and invasive biliary carcinoma.

Antibiotic and Chemotherapeutic Disease

Antibiotic and chemotherapeutic toxicity are frequently incorporated in fish diets and normally serious pathology is induced if excessive dosage or prolonged incorporation occurs. Erthromycin and sulfonamides are commonly associated with pathological effects. Specific toxic vacuolar degeneration of proximal of renal fibular epiltelisy. Retardation of growth venal tubular cadets, focal hepatic necrosis and visceral selerosid.

Binders

Use of polycellulose binders in high levels associated with a chronic degenerative condition lemony at hepatorenal syndrome. It is characterized by vacuolation, necrosis and ablation of renal tubules. The suspected complexes are metalothiol.

Standard artificial diets tend to eliminate these diseases where as substandard diets (home make, wet diet) used in supplementary feeding increase the possibility of their incidence.

Self Assessment Questions

1. What are nutritional diseases?
2. What are different types of Nutritional Diseases?

3. What are important Amino acid deficiency symptoms of fish and shellfish?
4. What are major Vitamin and Mineral deficiency symptoms of fish and shellfish?
5. What is blue, red and black gill disease?
6. What is sekoke disease?
7. What are SCP lesions?

Chapter 23

Probiotics and their Uses in Aqua Feeds

- ☆ Probiotic is an additive; alternative to antibiotics.
- ☆ Probiotics means "favour for life" (pro means favour or promote; bio means life).

Definition

Parker (1914) defines probiotic as "organisms and substances which contribute to intestinal microbial balance". The definition included microbial culture suspensions and crude microbial culture produces.

Probiotics are now defined by Prasad 1991 as, like microbial culture that are administered to animals with the primary aim of preventing infections diseases by strengthening the barrier function of the gut microflora or by non specific enhancement of their immune system.

Probiotics have also been defined as "organisms such as friendly bacteria, which contribute so much to the health and balance of the ecosystem in which it is involved".

Probiotics are live facultative beneficial bacteria when applied to shrimp pond, works in aerobic as well as anaerobic conditions, kills the harmful bacteria, digest the metabolites and reduce the organic load. In addition to keeping the environment congenial for the shrimp, these eco friendly bacteria enters the gut of shrimp through feed and fight against at the pathogens, restore the favourable microbial population, which in turn improve the feed utilization, enhance the host defense mechanism and promote faster growth.

In any normal healthy environment, there is a balance between the good or beneficial bacteria and the bad or detrimental bacteria. This is called as EUBIOSIS. During abnormal conditions or stress, a microbial imbalance is caused which is called DYSBIOSIS, which immediately leads to proliferation of harmful bacteria, disease condition and ultimately poor performance of the animals in the case of aquaculture causing heavy economic losses.

In Aquaculture, the actions of probiotics are as follows:

They are found useful in keeping the balance of the gut micro flora, which benefit the animal, by protection against disease and by improving nutrition.

They utilize nitrogen-bearing wastes in water and in the metabolism process, convert them into inorganic nutrients that can be easily utilized by the system.

It replaces or destroys harmful bacterial population by competitive inhibition. In addition, some of the antibacterial substances produced by certain bacteria, which has been selected, actively kill the harmful organisms mainly belonging to the Vibrio family including fungi.

High enzymatic activities such as amylase, protease, lipases, etc. remove organic load caused by excess feed, faecal matter and dead algae. The microbes create a perfect environment and balance the carbon nitrogen cycle.

In case of the soil also, the beneficial bacteria help in downgrading the harmful effects of the formation of sulphides and nitrites due to the heavy organic load accumulated during the culture period.

Probiotic is a Natural Alternative for Antibiotics

Antibiotics are commonly used for disease prevention in aquaculture. However, administration of antibiotics always have a negative effect by killing all bad and good bacteria in the system and there is a shift of natural balance from Gram positive bacteria to gram negative leading to DYSBIOSIS. In such conditions, animals cannot perform better and have more problems. Further to this, usage of antibiotics will lead to development of resistant pathogen. The probiotic limits the adverse effects of prophylactic antibiotic treatment in the cultured animal.

Probiotics products are available in the form of oral pastes, capsules, water dispersible powder, liquids and feed additives, which include microbial cell, microbial culture and microbial metabolites. Most of probiotics used to improve to feed utilization uptake of nutrients and to improve feed efficiency by altering rumen fermentation.

The microbial strains used as probiotics, should have the following characters.

Non pathogenic; Non toxic; Present as viable cells; Survival in the gut environment (resistance to low pH, organic acid etc.); Stable and viable for long periods under storage.

Both indigenous and imported probiotics are now available in the market.

Three different probiotics, which will be highly useful for the aquaculture farmers community for reaping better profits are 1. a gut probiotic (LACT-ACT) 2. a water probiotic (PRO-TECT), and 3. a soil probiotic (THIONIL). Either individually or in combinations, they will form the ideal weapons for a farmer to combat the microbial studies and improve the pond economy.

Commonly available probiotics contain Lactobacillus species like *L. acidophilus, L. factis, L. plantarum, Enterococcos faccacis* and yeast. The most commonly and extensive used one is *Lactobacillus.*

They are not only increase the growth rare of the fishes but also protect the fishes from the harmful microbes.

Potentials Benefits and Uses of Probiotic

- ✰ Breaks down the organic matter, unutilized feed and faecal matters.
- ✰ Enhances Nitrogen cycle; Prevents vibrio disease and other diseases.
- ✰ Probiotics used as a alternative to antibiotics; Detoxify poisonous gases like Ammonia and H_2S.

- ☆ Converts harmful Nitrite to Nitrate and Nitrates to Nitrogen.
- ☆ Increases the nutrient availability for plankton growth.
- ☆ Prevent formation of slime or sludge in pond bottom.
- ☆ Minimizes pond bottom acidity and activates the soil.
- ☆ Prevents depletion of dissolved oxygen level.
- ☆ Keeps water pH stable and balanced; increase average body weight of shrimp.
- ☆ Competes for food at microscopic level with dangerous microbes.
- ☆ Increases the appetite of shrimp; increase survival rate.
- ☆ Stimulates growth rate; improves FCR.
- ☆ Probiotics act as a alternative to antibiotics.

Self Assessment Questions

1. What are probiotics?
2. Compare Eubiosis and Dysbiosis.
3. Is probiotic is a natural alternative for antibiotics, if yes, justify.
4. What are types of probiotics?
5. List out the different actions of probiotics.
6. What are potentials benefits and uses of probiotics?

Chapter 24

Economics Aspects of Aqua Feed Usage

Feed Cost

It varies from one fish farming to other fish farming. In general, it will be 30 – 70 per cent even upto 80 per cent. For example; Carp farming 30 – 50 per cent; Shrimp farming 50 – 80 per cent. High feed cost is due to incorporation of higher per cent of protein usually the fish meal.

Cost of feed/kg of fish primarily depends on two factors.

Conversion Ratio of Feed in to Flesh

Conversion ratio can be reduced by eliminating waste and improving the feed formula.

Unit Price of Feed (Cost of feed ingredients)

Cost of feed could be reduced by improvement in conversion ratio or by minimizing the price of feed ingredients (or) combination of these two elements.

In livestock rearing, substantial savings of feed cost is possible by:

1. Reducing or minimizing physical wastage
2. Adopting proper feeding practices

General Potential way to Reduce Feed Related Cost

1. The use of appropriate feed types
2. Determination of the most cost effective ration size
3. Reduction of food waste

Efforts to Reduce the Cost of Feed in Aquaculture

1. Utilizing heteroprophic feed chain in aquaculture
2. Feeding fish with economically optimum level of protein
3. Adopting mixed feeding schedules
4. Using of growth promoters
5. Employing protein sparing ingredients
6. Replacing fish meal with non conventional feed ingredients (Silk worm – 100 per cent; fish silage – 100 per cent; Earth worm meal – partial; Soybean meal – partial.

Utilizing Micro-organisms and Heterotrophic Feed Chain

Micro-organisms as feed supplements to enhance the growth. Variety of algal, yeasts and bacteria has been employed as protein source. Microbial cell supply nucleic acids, proteins, vitamins, minerals and carotenoids for growth of fish. *Chlorella, Spirulina* are used as food for growing young fish. Astaxanthin is a major pigment present in *Paffia rhadozyma* alongwith ***b*** – Carotene, Zexanthin, lectin, when used in feed, fish accumulate more pigments in cells and which result in red colour. Promoting the growth of bacterial biofilm and zooplankton by providing organic substrates like sugarcane, baggasse, paddy straw and molasses is found to enhance fish production. Lab lab (BGA) is used in milk fish culture in South East Asian Countries like Philippines. SCPs rich in protein (40 – 70 per cent) on dry weight basis with generation time can be cultured using agricultural wastes and can replace fish meal in fish feed.

Mixed Feeding Schedules

Concept–2 to 3 days of high digestibility alternating with a day or two of low digestibility without lowering growth rate and yield. It also reduces the N_2 input to the system that prevents possible entrophication in ponds. Tilapia fry fed with high protein diet alternated with a low protein diet saved a feed cost up to 30 per cent. This new feeding schedule provides scope not only to reduce protein input, but also for efficient utilization of low quality proteins.

Feeding Fish with Economically Optimum Level of Protein

In optimum dietary protein level, maximum growth is obtained. Economically optimum level gives maximum economic returns.

De Silva *et al.* (1989); ↓ Tilapia

Zeitoun *et al.* (1976) ↓ Rainbow trout

↓

Optimum dietary protein level gave better results than economically optimum level, which result in substantial savings in feed cost.

Growth Promoters

Harvestable size achieved within short period without change in flesh quality. It saves on feed, labour and time. Examples: HCG (10 ppm), 17 MT (1ppm), Aqua gran (4.5 per cent), Amchemin AQ (5 per cent), Nutripro Aqua (1.5 per cent) for improving growth in carps, mahsheers and common carp.

Protein Sparing Feeds

Protein in the most expensive component in fish feed. In order to spare the bulk of the dietary protein, incorporation of other energy sources into fish feed will bring down feed the cost. Example: Lipids and CHO – efficient protein sparing ingredients.

Use of Non-conventional Feed Stuffs

By-product from Distilling Units

By-product from distilling units (26 – 70 per cent CP); trout (up to 21 per cent); channel cat fish (up to 40 per cent) including level; deficient in lysine; when supplemented with lysine, good growth.

By-product from Breweries

Dried breweries grains–good source of protein (19 per cent). No deleterious effect. Brewer's yeast – another protein source (40 – 45 per cent).

Starch Industries

Maize gluten and wheat gluten–by products from removal of most of starch from grains. Gluten (45 – 48 per cent CP). Tilapia (16 – 49 per cent) including level. Corn gluten poor in arginine and lysine, but good source of Zn, Fe, Niacin and vitamin C. Wheat gluten (12 per cent) incl. level.

By Product from Sugar Industries

Molasses (<3 per cent CP)

Limited use (1 – 9 per cent) including level, due to induction of fish stickness. At higher levels, it shows laxative effect. Upto 5 per cent level can be replaced with fish meal.

Press Mud

Included in traditional fish diet to a level of 25 per cent replacing rice bran. Press mud contains 15 per cent CP, 7 per cent oil apart from minerals.

By Product from Oil Industry

Oil cakes and meal – major by products from oil industries. Ground nut oil cake and soybean cake show better assimilation than caster seed cake, neem cake and sun flower cake.

By adopting the above strategies, either by single or combination, farmers can bring down the feed cost significantly which aids in sustainable growth of fish.

Self Assessment Questions

1. What are two factors considered in cost of production of feed?
2. List out general potential way to reduce feed related cost.
3. List out efforts to reduce the cost of feed in aquaculture.
4. What is mixed feeding schedule?

Chapter 25
Aquaculture Feed Management

Production of aqua feeds is, in fact, one of the fastest expanding agricultural industries in the world, with annual growth rates in excess of 30 per cent per year. Output increased from a level of 2.9 to 4.6 million tonnes between 1990 and 2000.

Production of 3 million tonnes of farmed marine/diadromous finfish/shellfish species (wet basis) in 1995 would have required over 1.5 million tonnes of fish meal and fish oil (dry basis) or the equivalent of over 5 million tonnes of pelagic fish (wet basis; assumes a pelagic to fish meal conversion factor of 5:1). This is not surprising as fish meal and fish oil usually constitute 50–75 per cent by weight of compound aqua feeds for most commercially farmed carnivorous finfish species and 25–50 per cent by weight (together with shrimp meals and squid meal) of compound aqua feeds for marine shrimp. Hence, many feed ingredient alternatives to fishmeal at varying levels are being sought in order to attain sustainable aquaculture in the current millennium.

The eventual success of these potential feed resources as fishmeal replacement in aqua feeds, however, will in turn depend upon the further development and use of improved techniques in feed processing/manufacture and feed formulation, including the increased use of specific feed additives such as feeding stimulants, free amino acids, feed enzymes, probiotics and immune-enhancers.

Feed Management in Fish Culture Systems

In developing appropriate feed management policies, numerous interrelated factors must be taken into consideration. These include economic, social, biological and environmental factors, which in practice range from feed selection to issues of final product quality as shown in following Figure.

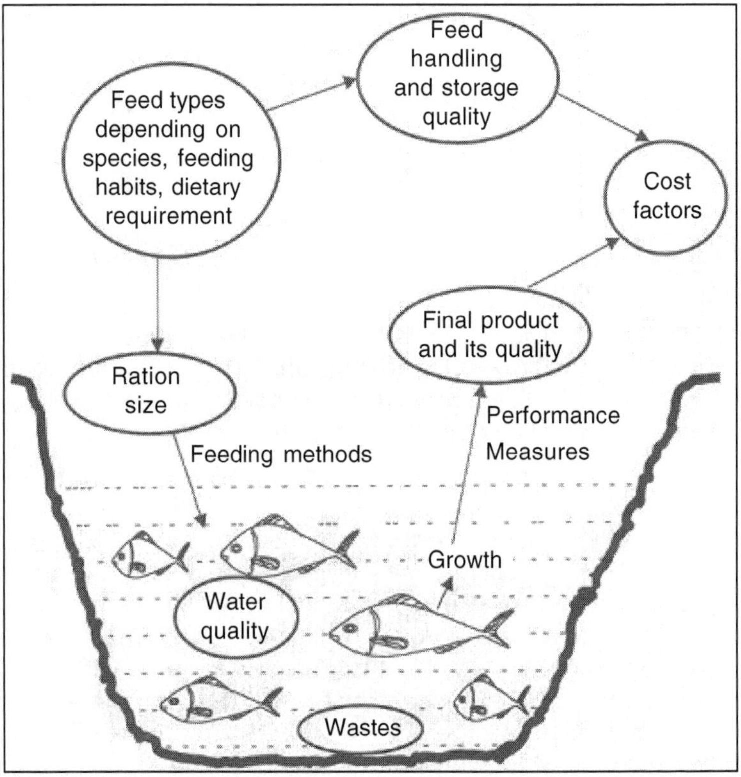

Diagrammatic Representation of Different Factors in Feed Management

Types of Feed

The foremost critical factor is selection of appropriate feeds and planning of optimal feeding regimens. Suitable feed should fulfill the nutritional requirements of species under culture. Proteins, lipids, carbohydrates, vitamins, minerals and water are the six major classes of nutrients, which are used for building, maintenance of tissues and supply of energy. The requirement for these nutrients varies depending on the species according to their feeding habit, habitat in which they live in and the stage in their life cycle. Our aim should therefore be to produce nutritionally balanced feed with optimum protein energy ratio. It should also ensure that nutrients are not lost in water during the feeding process. Therefore, aquaculture feeds of different formulations are processed using the special technologies to ensure the diet remains intact in water before ingestion, and that soluble nutrients are prevented from dissolving. These general categories of feeds used in aquaculture are wet feeds with moisture contents of 50–70 per cent, most formulated feeds with moisture contents of 20 – 40 per cent and dry pelleted feeds with moisture contents of less than 10 per cent. Since problems are associated with the distribution, handling, utilization, storage and quality of wet feeds and moist feeds, more and more dry feeds are manufactured either by steam pelleting or by extrusion pelleting. Advances in fish feeds and nutritional studies mean that many commercial feeds satisfying a wide range of options are now available.

Handling and Storage of Feeds

Optimizing handling and storage procedures on farms is an essential component of good management practice. High quality feed can readily spoil and denature if stored under inadequate conditions or for too long a period.

Incorrectly stored feeds may not only be unappetizing to fish or lacking in essential nutrients, but also may contain toxic and antinutritional factors. This can lead to abnormal behaviour, poor feeding response and growth. Hence, different feed types such as wet feeds, moist feeds and dry feeds must be handled and stored under appropriate conditions.

Feed ingredients and finished feeds should be labeled in accordance with national and international regulation so as to verify the quality of feed. A feed inventory should always be maintained including details of date of delivery, manufacturer, feed types, batch numbers, quantity, cost and any observation on the condition of feed on receipt.

Ration Size

The size of daily food ration, the frequency and timing of meals are the key factors influencing the growth and feed conversion.

Hence, the optimal feeding regimens must be determined as per the feeding behaviour, appetite and functioning of the digestive systems and the various specific chemical substances, which act as feeding stimulants for fishes.

Fish lose weight when their food intake falls below that required for maintenance. When ration size increases, the growth rate increases, up to maximum food intake and growth rate. Depending on the various biotic and environmental factors, food can be provided *ad libitum* (*i.e.* in excess) to apparent satiation or in restricted amounts.

Ration size is estimated by various methods using the feeding charts, feed equations, growth predictions and check tray etc. Besides the food ration size, the optimal food particle size also affects the growth and feed conversion efficiency. Large fish can ingest small particles, but it requires more energy to capture the required equivalent weight or smaller food particles. This results in measurable reduction in food conversion efficiency.

Attention should also be given to the influences of feed shapes, colours and textures of pellets on ingestion rates.

Feeding Methods

Production of high quality fish at least-cost depends on an effective feeding method. Various techniques exist, from hand feeding to mechanized feeding. They depend on diverse range of factors such as labour costs, scale of farming, species under farming, the type of holding system and hatchery or grow out systems. Often farmers use a combination of feeding methods such as hand feeding to mechanized feeding. A reliable and least-cost feeding system should ensure the effective distribution and spread of adequate feeds in aquaculture ponds.

Water Quality

The interrelationships between feeding and water quality in aquaculture is complex. By providing optimal species-specific requirements such as temperature, dissolved oxygen, pH and salinity, adequate feeding to satiation, improved growth and survival can be ensured.

When the water quality parameters fall below the optimal levels, feeding and growth will be impaired and the species under culture will be stressed. Accumulation of left over feed together with the excretory products is associated with high BOD, NH_3, H_2S, CH_4 and harmful effects of eutrophication. This is a critical issue in management since effluent quality can be linked directly to feeds and feeding practices and is regulated under water pollution control laws in many countries. Thus, feeding regimes should be designed to minimize the nutrient loss and faecal output and to maximize the nutrient retention and health status of the cultured fishes.

Measurement of performance

Effective feed management must be based on the assessment and interpretation of the data gathered by regular weighing. Survival, growth rates, feed conversion ratio and the product quality are the important parameters in any feed policy.

Cost Factors

Cost considerations in aquaculture feed management are vital in allowing farmers and entrepreneurs to select an effective feed so as to achieve a maximum profit with reduced FCR at a lower feed cost. The cost-effectiveness of feed is the most important factor from the various interrelated issues in sustainable aquaculture feed management, as illustrated in following figure.

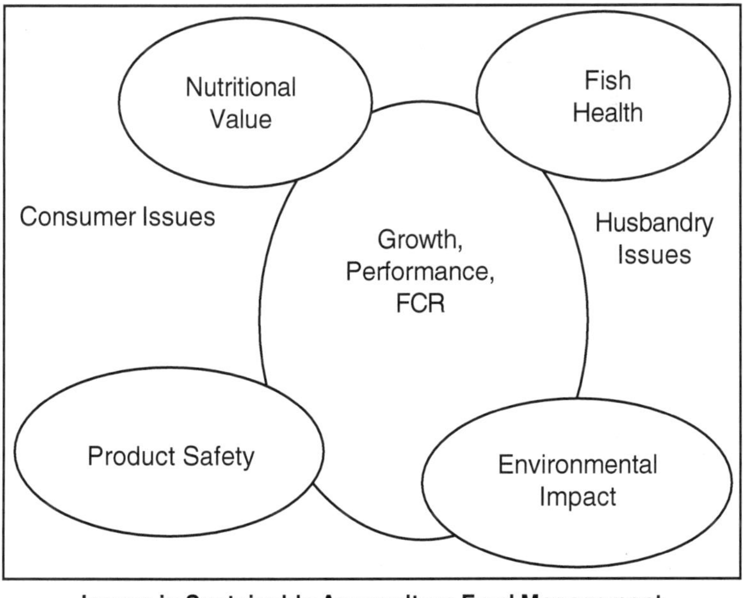

Issues in Sustainable Aquaculture Feed Management

Judicious feed management is an important factor in achieving good feed efficiency and reducing feed wastage. Selecting feeds, which are freshly prepared, quality assured and proven with best potential FCR, could reduce waste production. Poor quality and water stable feeds, which have lost their nutritional potency and are poorly accepted by the fish, should be rejected. Appropriate particle size of the feed should be designed for a particular stage. The ration size and feeding schedules should be regulated with reference to feeding guides, response of fish and environmental conditions.

Self Assessment Questions

1. List out numerous interrelated factors considered in developing appropriate feed management policies.
2. Provide diagrammatic representation of different factors in feed management.
3. Provide diagrammatic representation of issues in sustainable aquaculture feed management.

Index